MEMBRANE SCIENCE AND TECHNOLOGY

Industrial, Biological, and Waste Treatment Processes

MEMBRANE SCIENCE AND TECHNOLOGY

Industrial, Biological, and Waste Treatment Processes

Proceedings of a Symposium held at the
Columbus Laboratories of Battelle Memorial Institute
Columbus, Ohio, October 20-21, 1969

Edited by James E. Flinn
Chemical Process Development Division
Battelle Memorial Institute
Columbus, Ohio

PLENUM PRESS • NEW YORK–LONDON • 1970

Library of Congress Catalog Card Number 77-118126

ISBN-13: 978-1-4684-1853-8 e-ISBN-13: 978-1-4684-1851-4
DOI: 10.1007/978-1-4684-1851-4

Softcover reprint of the hardcover 1st edition 1970

A Division of Plenum Publishing Corporation
227 West 17th Street, New York, N.Y. 10011

United Kingdom edition published by Plenum Press, London
A Division of Plenum Publishing Company, Ltd.
Donington House, 30 Norfolk Street, London W.C.2, England

ERRATUM

In the paper "Separation of Blood Serum Proteins by Ultrafiltration" by Carel J. van Oss and Paul M. Bronson, the photographs in Figure 3 on page 143 and Figure 5 on page 146 have been accidentally interchanged. Thus, the caption on page 143 pertains to the illustration on page 146 and the caption on page 146 to the illustration on page 143.

Membrane Science and Technology (J. E. Flinn, ed.)
Plenum Press, 1970

PREFACE

This book is a collection of papers derived from a conference on membranes held at the Columbus Laboratories of Battelle Memorial Institute in Columbus, Ohio, on October 20 and 21, 1969.

When a decision is made to sponsor a membrane conference, the problem immediately arises as to what aspect of the technology needs to be emphasized. There were several alternatives from which to choose. The Office of Saline Water, for example, has been supporting for many years a tremendous volume of research on the desalination of sea and brackish water. In fact, were it not for this effort, the conference which resulted in this book could probably not have been held. Regardless, one could not easily choose to hold a conference on water desalting because the subject is adequately covered in the literature, and yearly conferences are sponsored by the funding agency. Other government agencies, specifically The National Heart and Lung Institutes and The National Institute of Arthritis and Metabolic Diseases, have supported a sizable number of research programs involving the use of membranes for biomedical devices useful in blood oxygenation and kidney augmentation or replacement. Again, these groups have their own outlets for disseminating research results. Still other choices existed among such areas as permeation processes for petroleum separations, advanced or novel membrane process concepts, or characterization of membranes – morphology, permeation properties, etc., – or biological membranes. None of these areas seemed to provide just the right technological emphasis.

The technological aspect actually selected was chosen with several thoughts in mind. First there was the knowledge of Battelle's interest and history of research in several areas relating to the membrane field, namely, membrane synthesis, polymer chemistry and technology, food product development and processing, biochemistry, biomedical engineering, and environmental waste management. Secondly, and perhaps more significant, was an intuititive feeling that membrane technology, as a processing tool, was beginning to pique the interest of many industries with separation problems of a common nature. Industries manufacturing food, fermentation, medicinal, and biological products are particularly pertinent examples and, indeed, representatives of these were in the majority at the conference. The commonality is in the use of membranes to fractionate, purify, or concentrate primarily aqueous solutions containing small dissolved ionic and neutral solutes **and** much higher molecular weight compounds, e.g., proteins, enzymes and viruses.

These types of systems present problems which are quite different to those encountered in simple water recovery from sea or brackish water. Cake buildup on the membrane surface, clogging of pores, adsorption in and on the membrane, rheological limitations on the mass transfer of solute and solvent due to viscosity, shear, and non-Newtonian fluid effects, are examples. Many of the papers in this volume deal with or allude to these problems and provide useful insights into their causes and preventions.

The papers are arranged in the same order in which they were presented. The first two, fundamental in nature, discuss and interpret phenomena relating to the transport of both large and small organic solutes within and through polymeric membranes. The next five papers are related in at least one aspect, namely, the application of membranes to solutions where macro-molecular compounds, principally of biological origin, must be separated, purified of dissolved ions or fractionated. As a group these papers are significant in that the unique problem of cake formation and membrane fouling due to solute retention at the membrane is clearly discerned and discussed from several viewpoints. Then follows two papers which are concerned solely with the transport of small dissolved ions for the production of chemicals and prospective hydrometallurgical separations. The final four papers describe the application of membrane processes to industrial water and waste treatment situations. Here, as in the processing of biologicals, both macro- and microsolutes are frequently constituents of the same solution. The recovery of water and the chemicals for recycle are the two needs which are stressed.

Since publication of this book represents the culmination of almost two years of effort on the planning and execution of a successful conference, it is appropriate to acknowledge with sincere gratitude the help and advice of various parties at Battelle-Columbus and elsewhere. I should mention first the 15 speakers at the conference and their co-authors, all but two of which have a paper in this volume. These missing two are H. P. Gregor of Columbia University and L. F. Ginnette of U.S.D.A.'s Western Division. Among my colleagues at Battelle, Robert H. Cherry, Jr., Charles W. Cooper, Manfred Luttinger, Gerald A. Grode, Richard D. Falb, Morton Fels, Herman Nack and Lester L. Hinshaw comprised the committee which planned and carried out the conference program and its many detailed arrangements. In addition Richard D. Falb, William T. McComis and Morton Fels served as session chairmen for the affair. Finally the Report and Graphic Arts Services group at Battelle contributed to the editing, styling, proofing, and typing of the manuscripts for publication.

James E. Flinn
Conference Chairman

CONTENTS

THE USE OF MEMBRANE DIFFUSION AS A TOOL FOR SEPARATING AND CHARACTERIZING NATURALLY OCCURRING POLYMERS

Lyman C. Craig
The Rockefeller University
New York, New York

The principle of a sieve for grading or separating very large particles from smaller ones seems too simple and obvious a procedure to merit discussion before a scientific audience. In contrast, however, when an attempt is made to sieve very small particles – those in the dimension range of relatively small molecules – an overall problem is presented which is well worth discussing before even the most sophisticated audience, because it touches on so many phenomena rather poorly understood yet so important to membrane technology and biochemistry. Perhaps a good way to begin a discussion of the sieving potential of membranes will be to consider some of the reasons a sieving process becomes so complicated with very small particles of molecular dimensions.

In the first place, with large particles, diffusion plays a relatively minor role, but in the case of molecules, it presents a parameter of major significance. When filtration is used to bring about the desired sieving of molecular-sized particles, relatively high pressures on one side of the filter must be used. The solvent moves through the filter under the applied pressure gradient with an effect on the filter difficult to evaluate clearly, but diffusion of the solute also takes place because of the concentration gradient. The effective pore size tends to change more or less in the course of the filtration due to some of the pores becoming filled with particles too large to pass or to changes in the gel structure due to stress on the membrane.

On the other hand, if simple dialysis is used to bring about the sieving process, a much more simple and reproducible process is to be considered. Here, the driving force can be restricted solely to diffusion of the solute under a concentration gradient, and even a very thin, relatively fragile and flexible membrane can be calibrated and repeatedly used for dozens of comparisons without change of its porosity. Solutes too large to pass the pores have no effect on the effective porosity of the membrane. It is for these reasons that we have concentrated on the use of dialysis for our quantitative studies rather than ultrafiltration. The practical importance of ultrafiltration as a procedure in biochemistry, however, is widely recognized. It often is an excellent complementary tool for the type of study under way in our laboratory.

In the second place, with large particles, the average particle diameter can be of the approximate order of magnitude of or larger than the thickness of the sieve. The size of the pores or holes in the sieve relative to the dimension of the particles can be quite accurately controlled. Here, filtration seems to be the ideal way to bring about sieving.

On the other hand, with particles of molecular diameters in the range of a few angstroms, a sieve of corresponding thickness is technically difficult to prepare and must be supported mechanically by a more porous material. Regulation of the pore size to the range required also presents a great technical difficulty. Nonetheless, progress has been made in overcoming these difficulties for certain solutes, notably in the desalination of seawater. Except for this type of skin filter, the lack of strength must be compensated for by making a sieve whose thickness is several orders of magnitude greater than the average particle diameter of the solutes. In this case, the solutes must follow a tortuous path through the sieve and surface effects are hereby greatly multiplied.

In the third place, the surface area of very small particles per unit quantity of substance is infinitely greater than for large particles. It is to be expected that surface interactions therefore will often play a much greater role, in fact, often the decisive role. It is, therefore, not surprising to find that, in dialysis, hydrophobic solutes require hydrophobic membranes and hydrophilic solutes require hydrophilic membranes.

In any case, in solutions where the particles are of molecular dimensions, the effect of molecular interactions must be one of the first considerations, since the dispersive action of the solvent itself involves a balance of molecular interactions. When the solute enters the membrane, this balance is altered more or less and the diffusive activity of the solute may be correspondingly altered.

Molecular interactions, except for rigid, porous materials, determine to a large extent the membrane structure. Let us consider cellophane as an example since it is one of the most widely used semipermeable membranes in biochemistry. Hydrated cellophane forms a sponge-like matrix, apparently because certain regions of the chain-like cellulose molecules interact with other regions, either of the same chain or of an adjoining chain, to form rather strong secondary bonds. Other regions are more strongly attracted to the solvent water molecules, and the porous structure thus arises which to the best of our knowledge behaves as if it consisted of a surprisingly narrow distribution of pore sizes.

It thus is obvious that the study of molecular interactions is of primary importance in molecular sieving not only from the standpoint of the solutes and solvent but from the membrane structure as well. Unfortunately, the methodology for studying molecular interactions is far from being as definitive as the membrane problem seems to need. Even the nature of the secondary forces which are involved are still a matter of speculation or theory. The literature speaks of van der Waals forces, hydrogen bonds, hydrophobic bonds, hydrophilic bonds, and charge-transfer bonds, but one frequently deals with structures which should permit several or all of these concepts to play a simultaneous role, and the contribution of each becomes largely a matter of opinion. It is interesting to note that carefully determined rates of dialysis can make a worthwhile contribution to our understanding of some of these forces, as will be seen later on.

Many approaches to the study of membranes have been employed in the past. Lack of time will not allow these to be discussed here. A more recent approach, developed in my laboratory, characterizes membranes almost entirely on the basis of the dialysis behavior of well-characterized pure solutes of known size and shape. This development(1), called "thin film dialysis", provides as nearly a maximum dialyzing area per ml of retentate as is experimentally practical. The dialysis cell shown in Figure 1 has been described in several publications(2 and 3) and will not be described here. A semilog plot of the percent of the total solute against time, as in Figure 2, allows interpretations to be made which, with known solutes, permits certain conclusions about the membrane to be made. Conversely, once a membrane has been properly calibrated, it can be used repeatedly to characterize an unknown solute with respect to its diffusional size and/or shape or the homogeneity of the preparation.

If a separation tool is to be used only for separation, it is not always necessary to know precisely the parameters which bring about the separation. On the other hand, if the operation offers promise as a tool for characterization, such as dialysis does, it becomes very important to know the parameters involved and why a certain degree of selectivity is achieved. Therefore, a first question to ask with membrane diffusion is whether or not the process is purely mechanical sieving or whether the membrane acts simply as a solvent barrier or both.

Data in favor of the former with cellophane, which indicate a separation on the basis of molecular size, are scattered through the literature. These data can be made much more convincing by use of the concept of restricted diffusion(4) as shown in Figure 3. Here, as the particle size approaches the size of the pore, the selectivity rapidly rises, as is indicated by the mathematical relationship shown in Figure 3. This effect can be expressed in another way by Staverman's concept of the reflection coefficient(5) as proposed in his studies with leaky osmotic pressure membranes. High selectivity, therefore, implies slow dialysis. This, however, can be largely overcome by the thin-film arrangement of Figure 1.

High selectivity also requires the availability of membranes whose pore size is carefully adjusted for the size of the particular solute being studied. This has not proven to be an objection, because we have found that wet cellophane tubing can be stretched(2) linearly to a certain degree, thereby causing its pore shape to be distorted. This reduces the effective pore size. On the other hand, two-dimensional stretching increases the pore size. The limits of alteration in terms of molecular diameters are about twofold in either direction and the effect is permanent and with sufficient stability for calibration. Further decrease in pore size can be achieved by partial filling of the pores by acetylation or further increase by zinc chloride treatment(6,2). Any desired pore size suitable for studying molecular sizes ranging from molecular weights of 18 to 100,000 can be achieved easily.

It is even practical to estimate how far to stretch a membrane to achieve a desired escape rate.(2) Thus, a membrane which originally gives the escape rate shown in Figure 4 for ribonuclease can be stretched about 20 percent in two dimensions. It will then give the escape rate shown for chymotrypsinogen. The Stokes radius of the larger protein is known to be about 25 percent larger than that of the smaller. This experiment offers convincing evidence that, here, the principle parameter controlling the rate of dialysis is mechanical sieving for the particular solvent, membrane, and conditions. If there were appreciable adsorption on the membrane, it could be recognized easily by removal of the solutions on each side of the membrane and determination of their solute content for a recovery comparison. A low recovery indicates adsorption on the membrane.

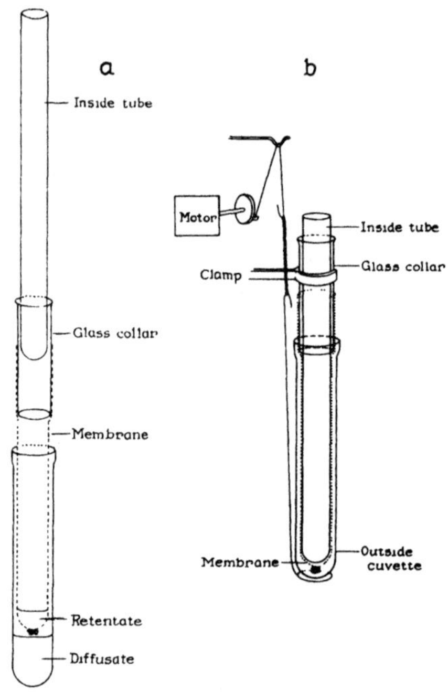

Figure 1. The analytical thin film dialysis cell.

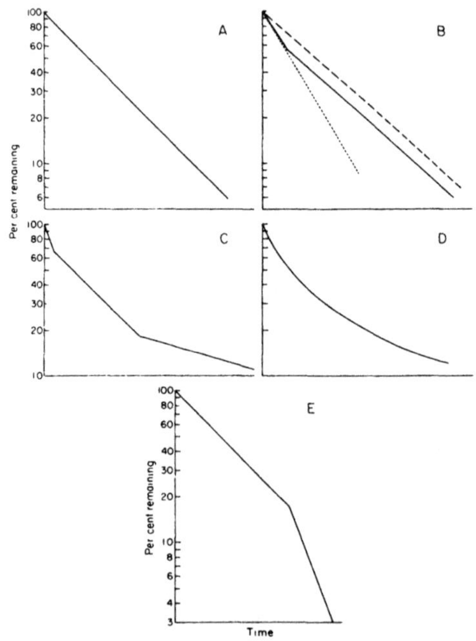

Figure 2. Plots used in the interpretation of thin film dialysis data.

$$A = A_0 (1 - r/R)^2$$

Figure 3. Schematic representation of restricted diffusion; A_0 = average cross-sectional area of hypothetical pore, A = effective cross-sectional area, γ = radius of particle, R = radius of pore.

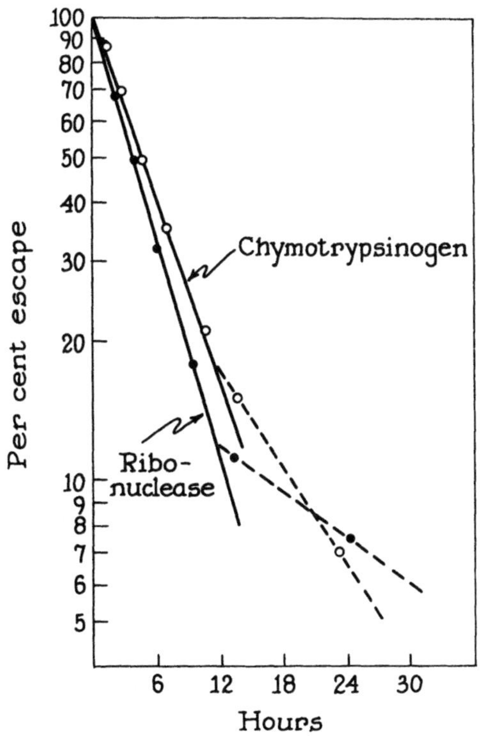

Figure 4. Experimental result correlating degree of stretching with change of pore size.

A rather precise indication of selectivity in terms of Stokes radii can be obtained by the use of suitable models. The Schardinger dextrins(7) can serve as excellent models, since their size and conformation have been well documented by X-ray crystallography. They are doughnut-shaped cyclic polymers of glucose. Cycloheptaamylose has a diameter approximately 11 percent larger than cyclohexaamylose. A cellophane membrane whose pores have been adjusted to a suitable size will give the data shown in Figure 5. Here, when the half escape time of cyclohexaamylose is 6 hr, that of cycloheptaamylose will be 12 hr and that of cyclooctaamylose will be 22 hr. From these data, we can say that we should be able to detect differences in Stokes radii of the order of 2-3 percent.

On the other hand, if the membrane is too porous, a low selectivity is observed, as is shown by the lower curve in Figure 5. Actually, it has been found that certain batches of cellophane do not show as high a selectivity as others, even when the porosity is adjusted to the half escape time shown in Figure 5. On theoretical grounds from the reasoning in Figure 3, a membrane with a very narrow range of pore sizes could be much more selective than one with a broader range. It therefore would seem that the selectivity of a membrane determined with well-characterized solutes of different sizes can give an indication of the homogeneity of the porous structure in a given membrane. The experimental data in Table 1 can serve as an example.

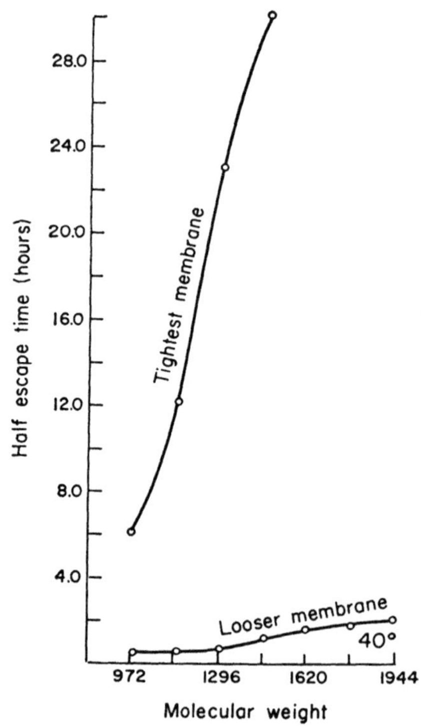

Figure 5. Demonstration of selectivity obtainable by thin-film dialysis with Schardinger dextrine.

TABLE 1. COMPARATIVE 50-PERCENT ESCAPE TIMES (HR) IN SELECTED CASINGS

Casing	Insulin	Lysozyme
36-DC-6	1.9	2.5
20-DC	3.5	5.6
36-DC-8B	4.1	36.0
36-DC-7	3.5	∞

The type of escape plot given can be a very useful tool for studying the solution properties of a given preparation(8) when it is determined in a membrane of high selectivity. An ideal pure solute always gives a straight-line or first-order escape plot. A mixture of two different sizes does not, as shown in the solid line of curve B in Figure 2. If three sizes are present, curve C could be obtained, or several could give a continuous curve as in D. A single solute that associates with concentration dependence can give a reverse curvature or break as shown in E of Figure 2. Depending on the concentration, temperature, and rate of association-dissociation, however, an aggregating solute can give types A, D, or E. Rerunning under different concentrations and temperatures is advisable to confirm the initial interpretation. The presence of a mixture can be confirmed and distinguished from association by isolation of the first and last diffusate and rerunning. When used this way, the chromatography (with Sephadex or Biogel) usually requires less labor, especially when more than two components are present. Thin-film dialysis, however, has an inherently higher selectivity and will reveal impurity in a preparation when gel filtration will not. If separations by dialysis are to be considered, the continuous form of thin-film dialysis(9) offers much more promise. This will be discussed further on.

Thin-film dialysis has been used as an effective tool for studying association-dissociation on more than one occasion. These include insulin(10), glucagon(11), hemoglobin(12), the tyrocidines(13), and certain dipeptides(14). In every case, the rate of dialysis increased with more dilute solutions and, except for hemoglobin, reverse curvature of the escape plot was noted for certain conditions. The rate of dialysis was highly sensitive to ionic strength and dissociating agents such as alcohol, dimethylformamide, or urea. Addition of small amounts of salt markedly slowed dialysis, but the dissociating agents had the opposite effect.

The tyrocidine system has been studied the most carefully. The tyrocidines are rigid polypeptides(13) in the molecular weight range of 1300. One of them, mistakenly called gramicidin-S, has now been studied carefully by high-resolution NMR and its conformation precisely established.(15) The sequence formulae are given in Figures 6 and 7. Gramicidin S-A in contrast to the tyrocidines A, B, and C behaves ideally and does not associate. It readily diffuses through a membrane in 0.1 N acetic acid to give a straight-line escape plot. The same membrane will not allow the tyrocidines to pass at all. A membrane of porosity suitable for studying proteins of molecular weight 6,000-10,000 is required for the tyrocidines. Alcohol increases the dialysis rate, as can be seen from the result in Figure 8. It is interesting to note that this dissociation phenomenon can be followed by NMR(16), as shown in Figure 9. In the associated form, an effect called "line broadening" prevents individual resonances from being

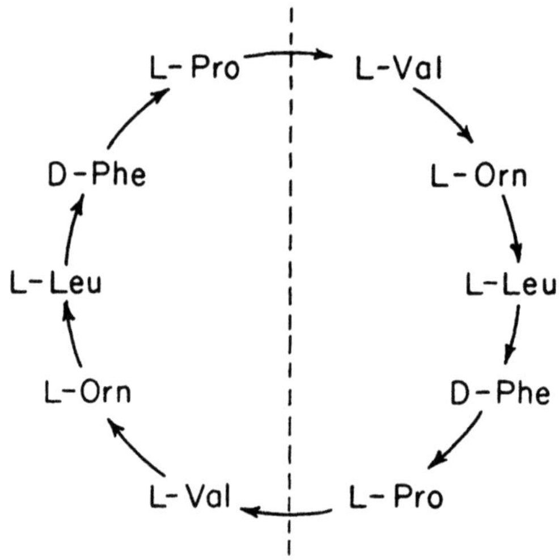

Figure 6. Amino acid sequence of gramicidin S-A.

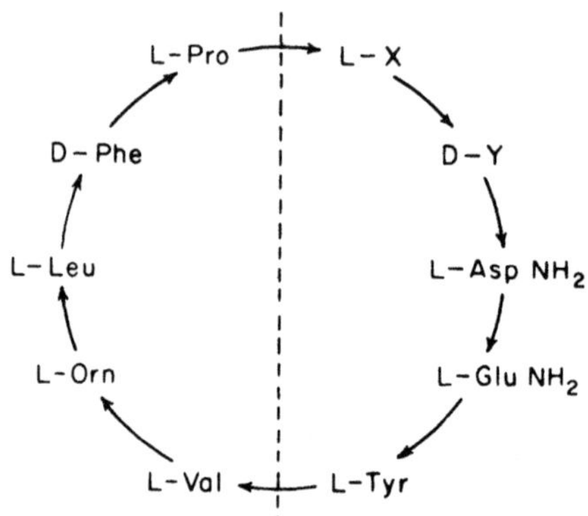

Tyrocidine	Residue X	Residue Y
A	Phe	Phe
B	Try	Phe
C	Try	Try

Figure 7. Amino acid sequences of tyrocidines A, B, and C.

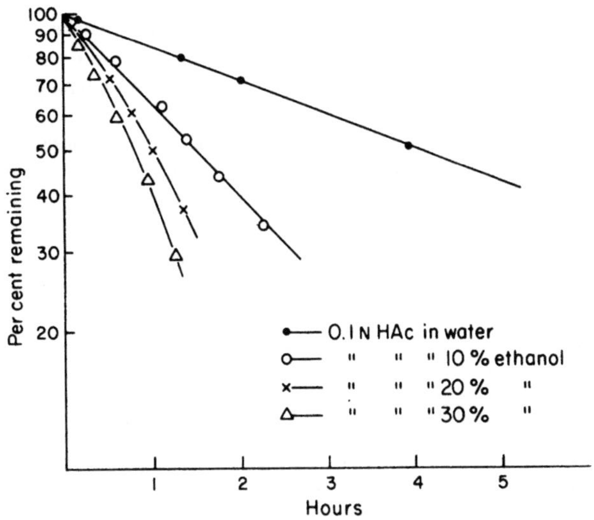

Figure 8. Effect of alcohol on the escape rate of tyrocidine B.

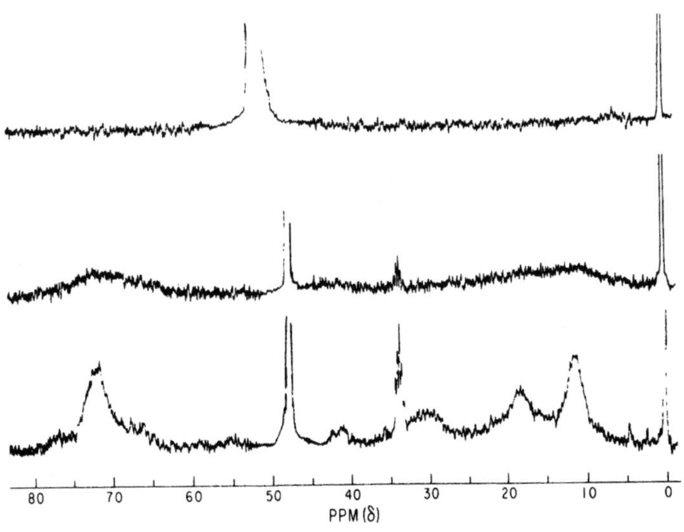

Figure 9. Effect of methanol on the NMR spectrum of tyrocidine B. Upper = deuterium oxide; middle = 20% deuterated methanol; lower = deuterated methanol.

observed. Methanol dissociates the aggregate and the expected resonances appear. This is a phenomenon similar to that observed when a native protein is denatured.[17] In the case of the protein, the effect is caused by intrachain but intramolecular interaction (tertiary structure), while it is intermolecular (quaternary structure) with the tyrocidines. In the latter case, a change of temperature or concentration produces a similar effect. Higher temperatures increase dialysis rates anomalously and reduce line broadening in the NMR spectra. Dilute solutions likewise give a faster rate of dialysis and a reduction in the line broadening.

It is known that polymeric substances such as proteins have a more or less rigid structure and a preferred conformation for that part of the molecule not rigidly held by cross-linking covalent bonds such as -S-S-bonds. Denaturing agents such as urea disturb the forces holding the molecule in the preferred conformation, and its shape can be changed drastically. This should be reflected in dialysis studies by considerably reduced rates of diffusion in a given membrane. This has been uniformly experienced[8] and contrasts strikingly with the behavior of solutes which aggregate but do not denature, such as the tyrocidines.

There are also many polymeric, naturally occurring substances such as polypeptides and nucleotides which have no covalent cross-linking stabilization and are therefore considered more or less random coils. It is to be expected that, in certain solvent environments, they would be more compact or globular than in others. It has been our experience that, with rigid, ideal solutes like gramicidin S-A, the solvent environment could be considerably changed by addition of alcohol, urea, salt, etc., or change of pH without significant change in the dialysis rate through a given membrane. This indicated that the porosity of the membrane was not significantly changed by the change of solvent. On the other hand, the so-called random coils or deformable molecules usually showed a marked change in the dialysis rate with even a slight change in the solvent environment.

The use of salt for studying the conformational flexibility of biopolymers by thin-film dialysis is particularly interesting.[18] Here, it seems to be the ionic strength that is the most important parameter involved, although in some cases there appears to be a specific ion effect due probably to ion binding. For example, in Table 2 is shown the change in dialysis rate for a series of proteins and polypeptides[8] brought about by addition of ammonium acetate to 0.15 M strength. The proteins, such as cytochrome C, are known by other methods to have relatively high conformational stability.

This salt effect seems to be due largely to the shielding effect on the charged groups[8], since it decreases the dialysis rate when the charges are both positive and negative along the chain but the reverse when the charges are of like sign, as in the case of the polynucleotides[18,19]. The effect is not exclusively due to charge, however. Addition of even small amounts of salt to a solution of low ionic strength changes the dielectric constant of the solution and can influence the hydrophobic bonding strength of certain groups which partly determine the conformation. It has long been known that salt fractionation can often bring about surprisingly selective separations and that a given ionic strength is required for optimum stability or enzymatic activity.

It has now been shown[7,20] by wide experience with many different types of solutes that comparative dialysis rates determined on the same membrane in the absence of adsorption effects can be interpreted in a manner similar to the data from free diffusion. Small differences in molecular size or shape are much more apparent because of the higher selectivity offered by restricted diffusion. Temperature coefficients should therefore be of considerable interest,

TABLE 2. CHANGE IN DIALYSIS RATE FOR SOME PROTEINS AND POLYPEPTIDES BY ADDITION OF AMMONIUM ACETATE TO 0.15 M STRENGTH

Peptide or Protein	Molecular Weight	Retardation Factor Due to Ammonium Acetate
Cytochrome C	13,000	1.1
Ribonuclease	13,600	2.5
Lysozyme	14,000	2.6
Cymotrypsinogen	25,000	6.0
Serum albumin	66,000	2.0
Salmiridine (a protamine)	4,000	6.0
Angiotensin	1,032	1.3
Glucagon	3,485	10.0
ACTH	4,540	10.0
Synthetic 23 residue ACTH	2,700	10.0
Parathormone	8,000	10.0

especially for the study of conformational changes and the phenomenon of association-dissociation. Before considering this possibility, it is of interest to investigate how closely the Stokes-Einstein relationship, $D = KT/\eta$, of free diffusion holds for thin-film dialysis where the half escape time in a given calibrated membrane is substituted for the diffusion coefficient, D. In general, it has been found that the relationship does hold to a first approximation with rigid ideal solutes such as gramicidin S-A which are not too highly hydrated. This is in striking contrast to solutes which are highly hydrated or known to have conformational flexibility or undergo association-dissociation. In studying the effect of a change of solvent environment, it is therefore important to measure the change of viscosity, also, and to make a correction according to the Stokes-Einstein relationship, as has been done in Table 3. These data show the effect of methanol on the comparative dialysis rates of several cyclic antibiotic peptides. As far as viscosity is concerned, our experience seems to indicate that such a correction is probably about 20 percent too high. This is not surprising, since the solution environment and viscosity

TABLE 3. HALF ESCAPE TIMES IN MINUTES

Peptide	Temperature	H_2O	50% MeOH	50% MeOH/H_2O
Gramicidin S-A	4°	13.0	21.0	0.92
	25°	4.8	8.0	0.98
	40°	4.0	5.5	0.75
Bacitracin	4°	15.0	70.0	2.4
	25°	10.0	26.0	1.5
	40°	9.0	12.0	0.75
Polymyxin B	4°	12.0	45.0	2.2
Tyrocidine B	25°	400.0	10.0	--

within the pores of the membrane are probably not the same as in free solution. Here again for the best evaluation we must rely on those model solutes such as gramicidin S-A, whose structure and properties are highly characterized and which conform as nearly as possible to ideality.

Finally, there is the question of fixed charge on the membrane. Obviously, for charged solutes such as the amino acids, nucleotides, and polypeptides, fixed charges on the membrane will make it impossible to correlate dialysis rates with diffusional size. In this case as in the case of adsorption[21], the rate can be greatly accelerated or retarded depending on the sign of the charge or the type of adsorption for the particular solute. For this reason we have searched for a membrane with minimum fixed charges and adsorptive properties. Visking dialysis tubing has been the membrane of choice. It has been shown to have an extremely low order of fixed charge and to have negligible adsorptive capability for most biochemicals.[21] In discussions of permeation rates in membranes, one often finds the idea that certain membranes "dissolve" the solute. This could be one of the forms of interaction of the solute with the membrane and, conceptually, would merge with the effect of adsorption.

With this rather brief coverage of some of the fundamental problems in dialysis, it is interesting to see if the information can be used to design a laboratory dialyzer with a much greater fractionation potential. A number of designs[22,23,24,25] with various claims are in the literature particularly, for use in hemodialysis. These will not be discussed here. A design from this laboratory based on the thin-film principle has been recently described.[26,27] It is not designed for hemodialysis but, rather, for laboratory separations and for analytical work. It is a continuous dialyzer which can be operated either cocurrently or countercurrently. The flow of the two streams is controlled by bilateral infusion pumps. The dialysis column (Figure 10) operates in a jacket whose temperature can be adjusted at will. The membrane is held in a thin annular space between two concentric tubes, with the outer tube rotating continuously. The rotation prevents channeling and keeps the membrane in microfluctuation. The details of the design and its evaluation have been described in full and will not be further treated here.

The arrangement provides the most rapid dialysis of a given solute through a given membrane we have yet been able to achieve. This can be tested by attaching the effluent tubes to fraction collectors. In this way, a pulse of a given solution of two solutes of different size injected into the retentate stream will provide complete separation provided one will pass the membrane but the other not. The effluent patterns will be of reproducible shape as in effluent chromatography. Figure 11 is a result obtained with a mixture of dextran blue and sodium chloride. Here, with a residence time of 6 min and with the diffusate stream run at about threefold the rate of the retentate stream, the salt is completely removed from the retentate stream. The performance can be more precisely evaluated with tritiated water and the use of a scintillation counter. Here, a pulse of tritiated water giving 10^8 cpm could be reduced to background count for a single pass. The dialyzer is thus an excellent tool for studying tritium-proton exchange.

It is also an excellent tool for studying the binding of various solutes and peptides to proteins. Even a peptide as large as bacitracin (mol. wt 1422) can be cleared to the extent of 97 percent on a single pass. By analysis of the effluent curves, a variety of molecular interactions are detectable, since molecular interactions influence diffusional activity.

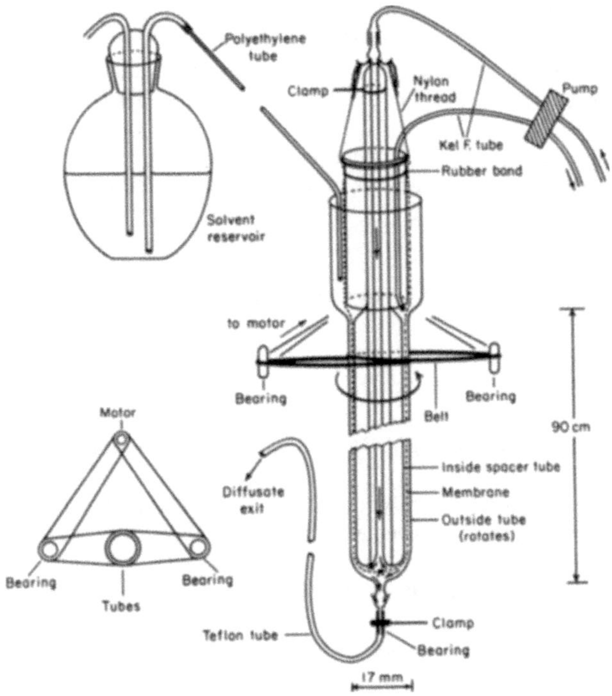

Figure 10. Schematic drawing of a thin-film countercurrent dialysis column.

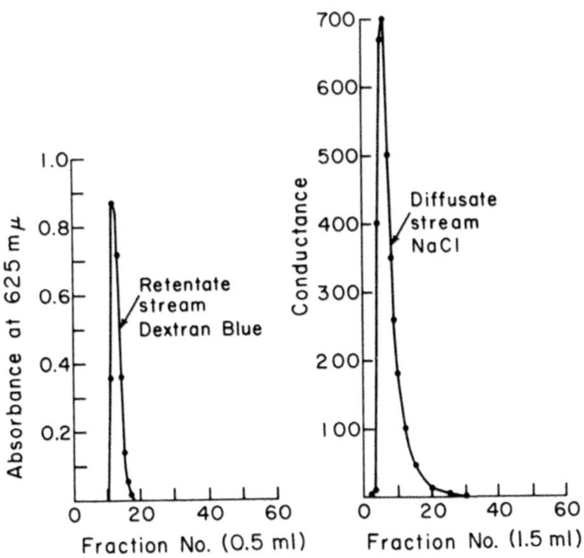

Figure 11. Effluent patterns from the continuous thin-film dialyzer.

We have not as yet completely evaluated the full potential of this tool, particularly, in separating closely related solutes. For this purpose, a number of tubes can be operated countercurrently in a train. This approach has shown a degree of promise for separating polynucleotides. These are particularly nonideal solutes which aggregate. Such dialysis trains permit the use of highly dilute solutions and thus reduce the tendency to interact.

When all points are considered, it seems likely that dialysis in the future will have a more important place in the experimental aspects of biochemistry than it has enjoyed in the past.

REFERENCES

(1) L. C. Craig, T. P. King, and A. Stracher, *J. Am. Chem. Soc.*, **79**, 3729 (1957).

(2) L. C. Craig and Wm. Konigsberg, *J. Phys. Chem.*, **65**, 166 (1961).

(3) L. C. Craig, "Advances in Analytical Chemistry and Instrumentation", Vol. 4, C. N. Reilley, ed., Interscience Publishers, New York New York, 1965, p 35.

(4) W. J. Ferry, *J. Gen. Physiol.*, **20**, 95 (1936).

(5) A. J. Staverman, *Trans. Faraday Soc.*, **48**, 176 (1948).

(6) J. W. McBain and R. F. Stucwer, *J. Phys. Chem.*, **40**, 1157 (1936).

(7) L. C. Craig and A. O. Pulley, *Biochemistry*, **1**, 89 (1962).

(8) L. C. Craig, "Methods in Enzymology", Vol. XI, C.H.W. Hirs, ed., Academic Press, New York, 1967, p 870.

(9) L. C. Craig and H. Chen, *Anal. Chem.*, **41**, 590 (1969).

(10) L. C. Craig, T. P. King, and Wm. Konigsberg, *Ann. N. Y. Acad. Sci.*, **88**, 571 (1960).

(11) L. C. Craig, J. D. Fisher, and T. P. King, *Biochemistry*, **4**, 311 (1965).

(12) G. Guidotti and L. C. Craig, *Proc. Nat. Acad. Sci.*, **50**, 46 (1963).

(13) M. A. Ruttenberg, T. P. King, and L. C. Craig, *Biochemistry*, **5** 2857 (1966).

(14) L. C. Craig, M. Burachik, and J. Chang, to be published.

(15) A. Stern, Wm. A Gibbons, and L. C. Craig, *Proc. Nat. Acad. Sci.*, **61**, 734 (1968).

(16) A. Stern and Wm. A. Gibbons, *J. Am. Chem. Soc.*, **91**, 2794 (1969).

(17) C. C. Macdonald and W. F. Phillips, *J. Am. Chem. Soc.*, **89**, 6332 (1967).

(18) J. Goldstein and L. C. Craig, *J. Am. Chem. Soc.*, **82**, 1833 (1960).

(19) L. C. Craig and W. I. Taylor, unpublished.

(20) L. C. Craig, *Science,* **144**, 1093 (1964).

(21) L. C. Craig and A. Ansevin, *Biochemistry*, **2**, 1268 (1963).

(22) R. Signer, H. Hanni, W. Noestler, W. Rottenberg, and P. von Tavel, *Helv. Chim. Acta*, **29**, 1984 (1946).

(23) H. A. Saroff and G.H.L. Dillard, *Arch. Biochem. Biophysics*, **37**, 340 (1952).

(24) B. J. Lipps, R. D. Steward, H. A. Perkins, G. W. Holmes, E. A. McLain, M. R. Rolfs, and P. D. Oja, *Trans. Am. Soc. Artif. Int. Organs*, XIII, 200 (1967).

(25) For discussions on the various types of hemodialyzers, see recent volumes of the American Society for Artificial Internal Organs.

(26) L. C. Craig and K. Stewart, *Biochemistry*, **4**, 2712 (1965).

(27) L. C. Craig and H. C. Chen, *Anal. Chem.*, **41**, 590 (1969).

ANOMALOUS TRANSPORT OF PENETRANTS IN POLYMERIC MEMBRANES

H. B. Hopfenberg
North Carolina State University
Raleigh, North Carolina 27607

INTRODUCTION

The fundamental study of small organic molecule transport in polymers has been actively pursued for more than two decades(1). These investigations have included a wide variety of polymers and penetrants over a broad range of experimental conditions. Many investigators were concerned with a specific transport feature which they implicitly suggested was characteristic of the given polymer-penetrant pair. These classes of behavior include:

Poiseuille or Pore Flow(2)
Concentration Independent Fickian Diffusion(3)
Concentration Dependent Fickian Diffusion(4,5,6,7)
Time Dependent Diffusion Anomalies(8,9,10)
Case II Transport(11,12,13,14,15,16)
Solvent Crazing-Stress Cracking(10,13,14,17,18,19,20)

Poiseuille or pore flow is observed only in isotropic microporous polymeric membranes where viscous flow is possible through the distribution of pores. All other modes of normal transport are diffusive whereby active transport is a consequence of a gradient of chemical potential of the diffusing species across the membrane. A third limiting case of transport involves polymer relaxation as the rate determining step rather than viscous flow or pure Fickian diffusion. When the relaxations are entirely controlling the transport kinetics, Case II transport is observed. The superposition of relaxation controlled transport and either of the other limiting transport mechanisms leads to time dependent anomalies.

The transport of small molecules in solid polymers is defined as Fickian when the transport satisfies one of Fick's equations for diffusion:

$$\frac{\partial c}{\partial t} = D \frac{\partial^2 c}{\partial x^2} \tag{1}$$

or

$$\frac{\partial c}{\partial t} = \frac{\partial}{\partial x} \left[D(c) \frac{\partial c}{\partial x} \right] \tag{2}$$

The following boundary conditions are conveniently met for Fickian sorption into a film:

$$c(x,0) = 0$$
$$c(0,t) = c(1,t) = C_o$$

where D is the diffusion coefficient, c the penetrant concentration, C_o the equilibrium concentration, t and x the independent variables time and distance, and l the film thickness. Due to the concentration dependence of the diffusivities for most organic vapor-polymer systems, Equation 2 normally applies for the diffusion of organic penetrants above Tg[1]. Normal or Fickian sorption is characterized by the following features:

(1) A linear relationship exists between the initial weight gain of the sample undergoing sorption and the square root of time.

(2) A smooth and continuous concentration profile exists through the film.

(3) The concentration becomes uniform across the film at long times.

Below Tg the transport process often becomes 'anomalous'. Frequently the diffusivity not only depends upon concentration, but upon time and position as well. Alfrey, Gurnee, and Lloyd[11] have proposed a simple limiting case for anomalous diffusion of organic vapors in polymers below Tg. Their Case II transport is characterized by the following features:

(1) At temperatures well below Tg, a linear relationship exists between the initial weight gain of a polymeric film undergoing sorption and time. (In contrast, Fickian sorption leads to a linear relationship between the initial weight gain and the square root of time.)

(2) A sharp boundary separates an inner glassy core of essentially zero penetrant concentration from an outer swollen, rubbery shell of uniform concentration (not a sufficient criterion for non-Fickian diffusion since sharp advancing boundaries have been observed for Fickian diffusion with a strongly concentration-dependent diffusivity).

(3) The boundary advances at a constant velocity.

This model, which Alfrey developed for a simple limiting case for anomalous diffusion, fits several of the anomalous features of the transport process below Tg observed by King and others. King[21] reported a linear relationship between the initial weight gain and time in his study of the sorption of alcohol vapors in wool and keratin. His explanation for the linear advance was a concentration-dependent diffusion coefficient which led to a build-up of a steep front which then moved through the medium. Other authors have reported anomalous behavior in their studies of organic vapor transport in glassy polymeric systems, but have offered no simple explanation for the behavior. Crank[22] did develop a model to describe the behavior with a time-dependent diffusion coefficient and/or a nonconstant boundary condition, $C_o(t)$; but in each case he had to include several additional parameters to satisfactorily describe the behavior.

Michaels, Bixler, and Hopfenberg[13] have also reported Case II transport in their study of the penetration of liquid n-heptane in polystyrene. This penetration led to solvent crazing of the film, which is an extreme example of Case II transport. They noted a sharp boundary between the crazed outer surfaces and the unaffected central core advancing with a constant velocity through a film. A study of the temperature dependence of this process revealed an apparent activation energy of approximately 60 kcal/g-mole, which is in the range of activation energies for stress relaxation of polystyrene and well above that for a normal diffusion process. More importantly, this activation energy is sufficiently high to reflect primary bond breakage and free-radical formation as a consequence of the advancing penetrant front. They concluded that the rate-controlling step of this transport process is the osmotically induced relaxation of the polymer at the boundary between the crazed outer layers and the unaffected central core. Diffusion to the boundary is rapid and does not affect the observed transport kinetics.

Since Case II transport is a relaxation controlled transport process, the parameters affecting relaxation such as polymer orientation, molecular weight, molecular weight distribution, temperature, penetrant activity, and penetrant physicochemical properties appear to be the important parameters to study in an investigation of the transport of organic molecules in glassy polymers. The effects of temperature, penetrant activity and penetrant structure on the sorption kinetics and equilibria of hydrocarbons in biaxially oriented and cast-annealed polystyrene have been presented by Hopfenberg, Holley and Stannett[14]. The effects of molecular weight and subtle orientation effects were presented by Bray and Hopfenberg[23]. A survey and generalization of the various transport features observed for alkanes in polystyrene was presented by Hopfenberg and Frisch[24]. This review summarizes their salient experimental techniques, results, and conclusions in an attempt to present a concise and up-to-date treatise on relaxation-controlled transport in polymers.

REVIEW OF EXPERIMENTAL TECHNIQUES USED IN VAPOR-SORPTION AND SOLVENT-CRAZING STUDIES

Vapor-Sorption Studies

Quartz helical springs (Worden Quartz Products, Incorporated, of Houston, Texas) with sensitivities of 1.95 to 1.97 mm/mg were used to measure weight changes due to sorption or desorption of the hydrocarbons in biaxially oriented polystyrene.* The spring extension was measured with an optical reader supplied by Misco Scientific, Berkeley, California. The precision of the relative weight measurement, determined by this extremely sensitive cathetometer, was ±2 μg.

*TRYCITE 1000. Dow Chemical Company, Midland, Michigan.

For a sorption run, a fixed amount of hydrocarbon sufficient to maintain a specified partial pressure was bled into a temperature-controlled (±0.1 C), evacuated spring case. The consequent spring extension was measured as a function of time to determine the sorption kinetics. Similarly, for a desorption run, the spring case was rapidly evacuated and readings resumed as a function of time. The temperature was varied from 25 to 50 C and the vapor pressure of the several penetrants varied over a range corresponding to activities of 0.38 to 0.89.

Liquid Sorption-Solvent Crazing

The polystyrene film sample was immersed in the pure normal hydrocarbon overlaying a pool of mercury for a specified period of time at a fixed temperature. The samples were removed from the solvent and placed in an isooctane bath at 0 C for 15 min to exchange the normal hydrocarbon and then allowed to dry at ambient laboratory conditions. The kinetics of the liquid penetration were studied by observing the advance of the boundary between the crazed and uncrazed film by examining microtomed cross-sections of the film with a light microscope.

RESULTS AND DISCUSSION

Qualitative Effects of Penetrant Sorption

The biaxially oriented polystyrene undergoing sorption of normal hydrocarbons changes from a transparent flat film to a curled film; the amount of curl increased with time and with increasing penetrant activity. The curl of the film reflects asymmetry in the thickness direction of the film specimen. This asymmetry is not unexpected in a commercially extruded film since the processing undoubtedly results in a nonuniformity of polymer microstructure.

At very high activities the osmotically induced stresses created by the penetrating molecules become sufficient to cause crazing of the film. The critical concentration at which crazing occurs was found to lie between 8 and 10 percent by weight of penetrant. Crazing of the film reveals that the osmotically induced stresses exceed the local tensile strength of the polymer, which results in microfailure. It is quite possible that crazing involves primary bond breakage as well as simple chain separation[19].

At the high activities, which resulted in crazing, shrinkage in the plane of the film occurs at long times which correspond to the penetrant advance to the film midplane. Crazed films appeared white and opalescent. Photomicrographs of crazed film cross-sections revealed an extremely sharp boundary between crazed and unswollen polymer (Figure 1)[12]. Shrinkage of the film is a manifestation of the relaxation of the stresses frozen into the film by orientation. Film shrinkage did not occur at the low vapor activities, since an insufficient amount of penetrant was present to accelerate long range polymer relaxations.

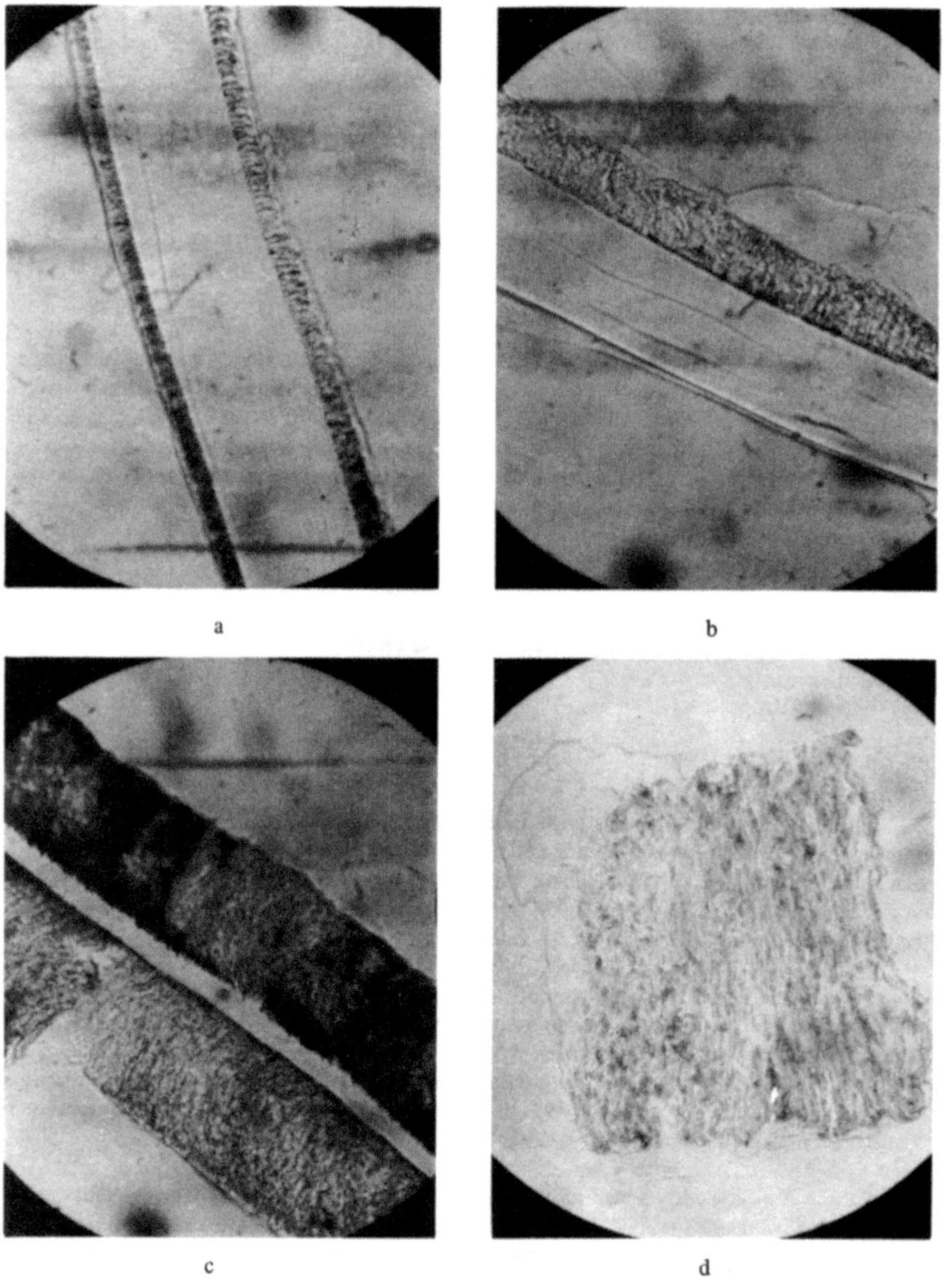

Figure 1. Cross-sections of 2 mil TRYCITE (biaxially oriented) film treated with n-heptane (magnification: 250X): (a) treated on both sides at 35°C for 20 min; (b) treated one side at 40°C for 15 min; (c) treated both sides at 35°C for 92 min; (d) treated on both sides at 40°C for 25 min.

Liquid Sorption-Solvent Crazing. The sorption of normal hydrocarbons in biaxially oriented polystyrene causes crazing of the film. This crazing is conveniently viewed as a simple extension of Case II sorption. If the osmotic stresses generated as a consequence of sorption are insufficient to cause local fracture, rate controlling polymer relaxations may still be biased by these stresses. Michaels, Bixler, and Hopfenberg(12) describe the appearance of n-heptane-crazed biaxially oriented polystyrene film as a white, opaque membrane which eventually shrinks in the planar directions and swells perpendicular to the plane of the film. This behavior was observed for sorption of a series of straight-chain hydrocarbons (n-pentane to n-nonane). The rate of crazing decreases as the molecular weight of the hydrocarbon penetrant increases. The crazing kinetics for the various penetrants at 35.5 C are presented graphically in Figure 2. The osmotically induced stresses at the advancing boundary between the crazed surfaces and glassy core, which are proportional to molar penetrant concentration, control the rate of polymer relaxation and consequent crazing rate. The molar concentration of penetrant decreases with molecular weight; therefore, a smaller osmotic stress is developed for sorption of the higher homologs at equal gravimetric sorption levels.

The craze-front velocity, characterized microscopically, is a measure of the rate of sorption. The craze front velocities are given in Table 1. They were observed to be constant with time for a fixed temperature over the range of temperatures studies for each penetrant. By studying the temperature dependence of the craze-front velocity for the three penetrants investigated, apparent activation energies for the penetrants are obtained as follows: n-pentane, 50 kcal/g-mole; n-hexane, 42 kcal/g-mole; and n-heptane, 38 kcal/g-mole. These Arrhenius plots are given in Figure 3; the activation energies are tabulated in Table 1. The molar concentration of penetrant increases from n-heptane to n-hexane to n-pentane. As the molar concentration of penetrant increases, the osmotic stress is increased. In addition, the swollen polymer is more

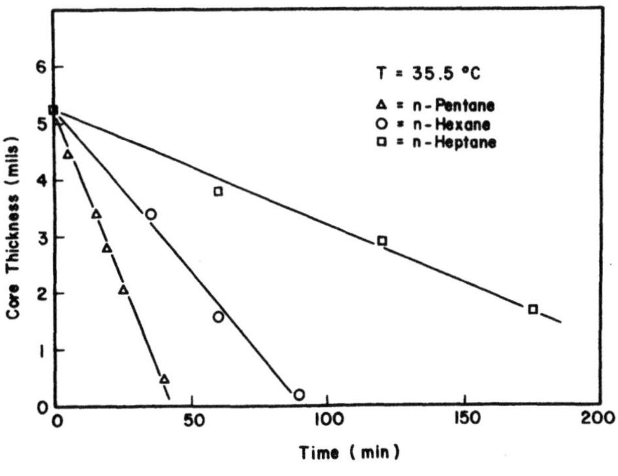

Figure 2. Dimensions of residual, uncrazed core as a function of time for biaxially oriented polystyrene immersed in various liquid normal alkanes at constant temperature.

TABLE 1. RATES OF CRAZE FRONT ADVANCE[(a)] IN GLASSY POLYSTYRENE

Liquid	Temperature, C						Activation
Penetrant	25.3	30.0	35.5	40.0	45.0	50.0	Energy, kcal/g-mole
n-pentane	0.675	2.55	9.75	--	--	--	50
n-hexane	--	0.85	2.86	8.30	22.5	--	42
n-heptane	--	--	1.00	2.15	6.85	14.5	38

(a) (mils/min) x 10^2.

plasticized at the higher penetrant concentrations. It is reasonable to assume, therefore, that larger polymer segments cooperate in the relaxation process during sorption of the lower homologs which are sorbed to a higher concentration. The increased activation energy for transport of the lower homologs appears related to the segment size involved in the relaxation process controlling transport of these penetrants. The value of 38 kcal/g-mole for n-heptane crazing is significantly lower than the 60 kcal/g-mole reported previously for n-heptane crazing of polystyrene(12). Due to the probable morphological differences between the two polystyrene films acquired 4 years apart and the suspected dependence of transport mechanism on molecular weight, molecular weight distribution, and polymer orientation, a difference in the activation energies is reasonable. However, each value is in the range for stress relaxation of polystyrene and indicates a relaxation-controlled transport process for liquid sorption. These high activation energies also suggest primary bond breakage leading to free radical formation. Experiments are presently being run to check the occurrence of this phenomenon.

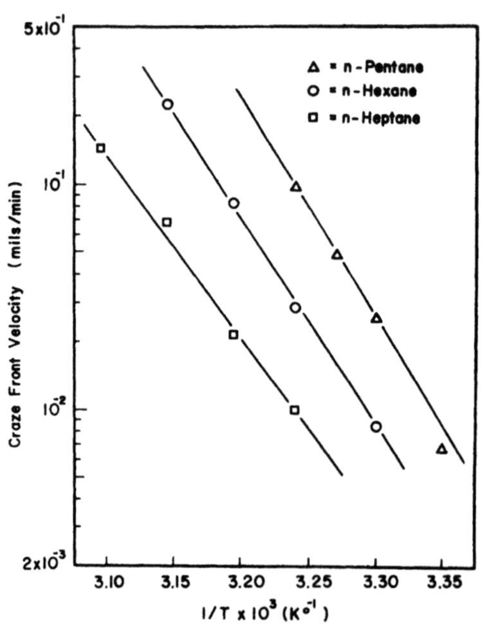

Figure 3. Arrhenius plot of craze front velocity as a function of temperature.

The apparent activation energy of 50 kcal/g-mole for liquid sorption of n-pentane is considerably higher than the 28 kcal/g-mole calculated for the vapor sorption process at an activity of 0.63. The increased activation energy for the liquid sorption process (unit activity) is consistent with the larger osmotic stresses which are developed by the higher equilibrium molar concentration in the liquid sorption. These larger stresses result in cooperative motion of larger segments of the polymer molecules and, therefore, a higher activation energy although the same basic kinetic mechanism (sorption linear with time-Case II transport) controls transport at activities of 1.0 and 0.63.

Sorption of liquid isopentane in the polystyrene film revealed none of the features shown by the straight-chain hydrocarbons. This result led to the conclusion that the branched hydrocarbon sorbs to an insufficient molar concentration in the polymer matrix to induce solvent crazing.

Effect of Temperature on Vapor Sorption

The results of a series of vapor sorption studies at a constant vapor activity of 0.63 over a temperature range of 25 to 40 C are presented in Figure 4 and Figure 5. The equilibrium concentrations measured at 0.63 vapor activity are nearly identical over the range of temperatures and, therefore, indicate only a small heat of mixing for the n-pentane-polystyrene system. The weight gain of n-pentane is linear with time, but the curves do become slightly concave toward the time axis at long times, suggesting a possible Fickian contribution to the Case II transport process at long times. Since the sorption curves have nearly the same equilibrium concentration, an apparent activation energy can be calculated from the temperature dependence of the initial sorption rates. The calculated activation energy is 28 kcal/g-mole. The value is much higher than that for normal (Fickian) sorption. The similarity of the shape of the sorption versus time curves over the temperature range studied indicates that the mechanism for sorption over this range of temperatures at an activity of 0.63 is invariant.

Effect of Penetrant Activity on the Sorption Process

A series of sorption experiments were run at a temperature of 30 C over a range of penetrant activity from 0.45 to 0.89. The sorption equilibria are described by the isotherm of Figure 6. The sorption kinetics are presented in Figure 7. A definite change in mechanism from Case II sorption towards Fickian sorption occurs with decreasing vapor activity. The curves which show a linear time dependence to sorption equilibrium suggest that the penetrant advances behind a boundary moving with a constant velocity separating the outer shell and glassy core. This boundary moves from the surface to the center and then abruptly stops. Presumably, an equilibrium concentration exists behind the advancing penetrant front.

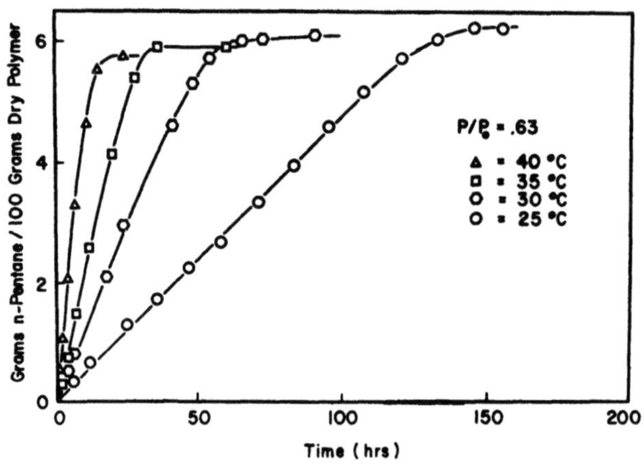

Figure 4. Kinetics of n-pentane sorption in biaxially oriented polystyrene at constant activity over a range of temperatures.

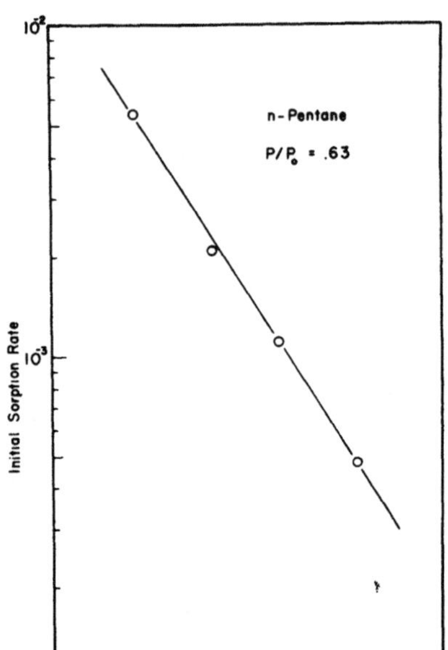

Figure 5. Arrhenius plot of n-pentane sorption rate at constant activity as a function of temperature.

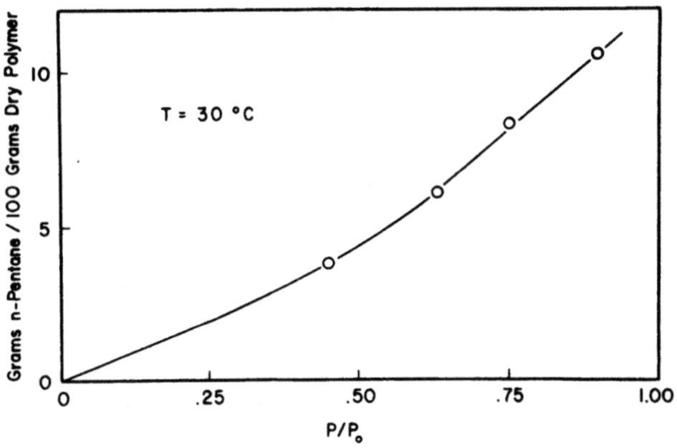

Figure 6. Equilibrium sorption isotherm for n-pentane sorption in biaxially oriented polystyrene at 30 C.

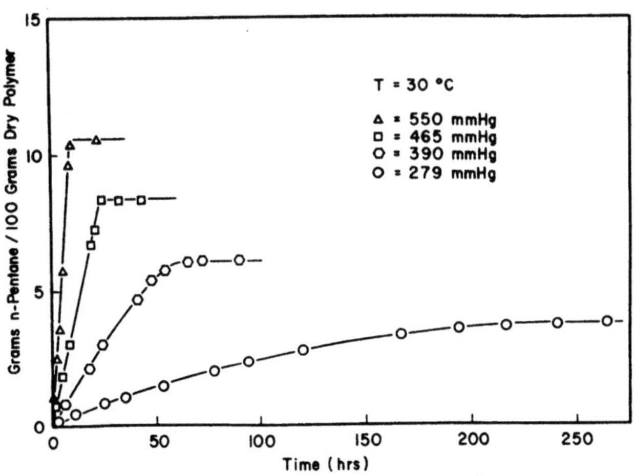

Figure 7. Kinetics of n-pentane sorption in biaxially oriented polystyrene at constant temperature over a range of n-pentane activities.

Kinetics of Penetrant Desorption

Characterization of the n-pentane sorption kinetics at activities less than 0.45 are prohibitive since the experimental time for each run becomes the order of months. The activity regime below 0.45 was studied by monitoring the desorption kinetics of samples equilibrated at an activity of 0.45. Film shrinkage did not occur in these samples as a consequence of sorption, hence the desorption process indeed occurred in a biaxially oriented sample.

Although the sorption of n-pentane in biaxially oriented polystyrene was dramatically non-Fickian for most of the conditions studied, the desorption behavior in all cases was very nearly Fickian. The most reasonable means for characterizing the kinetics of desorption was, therefore, the calculation of diffusion coefficients. The diffusivities were calculated using the long-time method to obtain a value of the diffusivity at zero penetrant concentration, D_o(13).

Study of the kinetics of desorption of n-pentane into a dynamic vacuum of 10^{-6} mm Hg over a range of temperatures permits evaluation of the activation energy for transport at a concentration approaching zero. This study was performed for n-pentane in biaxially oriented polystyrene and the results are summarized in the Arrhenius plot of Figure 8. The slope of the plot corresponds to an activation energy for diffusion of 5.6 kcal/g-mole.

A rather interesting pattern emerges if one analyzes these data in comparison with the liquid sorption studies (unit activity) and the vapor sorption studies conducted at an activity of 0.63. As activity is reduced, the temperature dependence of the sorption kinetics is characterized by a decreasing activation energy. The activation energies at activities of 1.0, 0.63 and

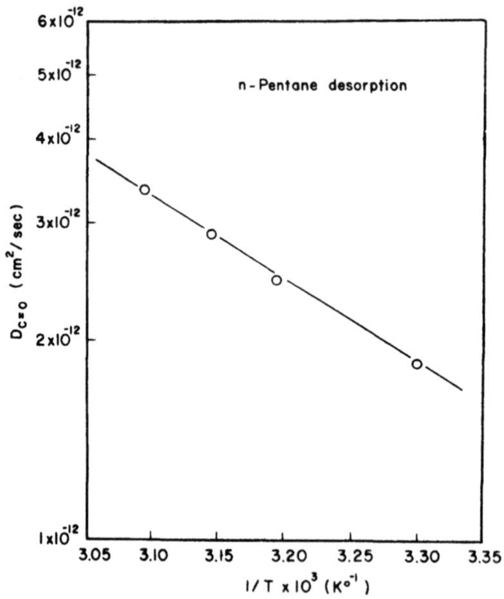

Figure 8. Arrhenius plot of n-pentane diffusion coefficient at zero concentration as a function of temperature.

zero are 50, 28, 5.6 kcal/g-mole, respectively. The activation energy at zero penetrant concentration is consistent with a sorption mechanism completely controlled by Fickian diffusion.

At unit activity, the sorption uniformly follows Case II transport including solvent crazing. As activity is reduced to 0.63, crazing is no longer observed although Case II sorption is still evident with a small Fickian contribution. The kinetics of penetrant transport approaching zero activity are largely described by Fickian diffusion. The activation energy characterizing this Fickian transport, 5.6 kcal/g-mole, is low but in the range of normal diffusion processes. Since the relaxation processes responsible for Case II sorption are biased by the osmotic stress which is monotonically related to penetrant concentration, the suggested shifts in mechanism for transport seem quite explicable. Simply, as concentration is reduced, the relaxation processes are markedly diminished. The absolute rates of transport are exceedingly low at reduced activities; the diffusion coefficients for the desorption process are in the range of 10^{-12} sq cm/sec.

A more thorough study of the effect of activity on the activation energy for transport at activities below ca. 0.63 becomes prohibitive since the experimental time for each run becomes the order of a month. Since the trends reported here are wholly consistent, it appears that the generalizations suggested with respect to the shift in transport mechanism with changing penetrant activity are indeed valid.

Generalized Transport Behavior

The transport features observed for normal hydrocarbons in polystyrene are qualitatively quite similar to those observed in other, rather diverse, polymer-penetrant systems[15,16]. The similarities in qualitative behavior suggest that the diverse behavioral features noted in polystyrene probably occur for most amorphous systems if a sufficient range of temperature and activity (traversing the glass transition range) is encompassed by experimental conditions. These patterns of behavior seem quite diverse and complicated when listed as above. The relationship between the various transport features are more easily assimilated, however, by examining the various regions of the temperature-activity plane presented in Figure 9.

The following features are readily apparent:

(1) Time dependent anomalies and Case II sorption including solvent crazing are confined to relatively high penetrant activities and temperatures in the vicinity of and below the effective Tg of the system. The effective Tg of the system is represented by the dashed line extending through the anomalous diffusion region.

(2) The region of Case II sorption (relaxation-controlled transport) is separated from the Fickian diffusion region by a region where both mechanisms are operative giving rise to diffusional anomalies.

(3) The activation energy characterizing Case II sorption decreases as penetrant activity is reduced.

(4) Concentration-independent diffusion is only apparent at very low temperatures and/or activities.

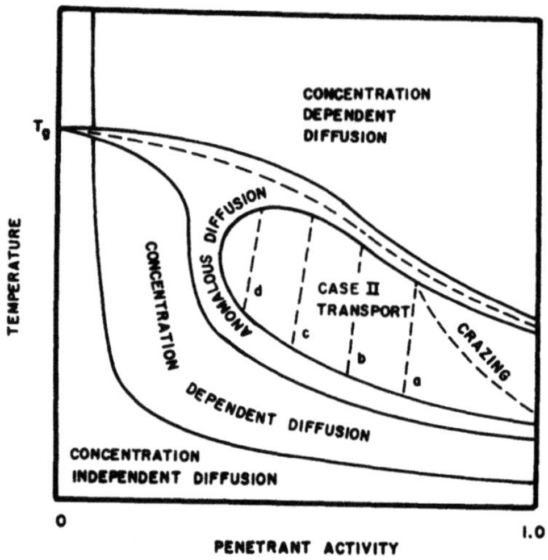

Figure 9. Transport mechanisms for n-pentane in biaxially oriented polystyrene as a function of temperature and penetrant activity. ($E_{ACT_a} > E_{ACT_b} > E_{ACT_d}$)

It should now be readily apparent that diverse transport behavior occurs for a specific polymer-penetrant pair if the investigator is willing to extend his analyses to a broad range of temperature and penetrant activity. One must specify a limited temperature-activity regime when they refer to the likelihood of observing a given transport feature (viz. crazing, Case II transport, anomalous diffusion, etc.). All of the limiting and intermediate transport features can most likely be observed experimentally if a sufficiently broad temperature-activity plane is explored. For some systems, insufficient penetrant levels may be achieved at equilibrium to generate an osmotic stress large enough to bias relaxations controlling Case II transport and/or solvent crazing. For these special systems, the crazing region and possibly the Case II transport regime would disappear from the temperature-activity plane.

Work in progress in our laboratory suggests that subtle changes in molecular weight, molecular weight distribution, or orientation will have rather pronounced effects on the quantitative position of the various regimes(23). To be completely general, additional temperature-activity planes for each independent variable of the polymer (for a given penetrant) would have to be constructed.

Molecular Weight Effects

Since the evidence has so overwhelmingly suggested that Case II transport including solvent crazing is controlled kinetically by relaxation processes, it appeared rather challenging to determine whether the variables associated with the polymer molecules which affect relaxation processes do, in turn, affect Case II transport and solvent crazing. Polymer molecular weight and polymer orientation affect the rates of relaxation[18,23]. The objective of this study, therefore, was to determine the effect of molecular weight and subtle changes in orientation on the rate of solvent crazing of polystyrene by normal pentane.

In 1963 Rudd[18] measured the rate of relaxation of polystyrene samples of molecular weight 100,000 to 300,000 exposed to butanol at unit activity under an initial stress of 1000 psi. Relaxation rates dropped by a factor of 10^6 with increasing molecular weight over this rather narrow range of molecular weights which was related to greater hindrance to chain uncoiling and viscous flow in the higher molecular weight samples.

Hopfenberg[20], however, found that film cast from polystyrene of molecular weight 240,000 crazed faster than that of weight 180,000. These films were cast on glass and, therefore, developed a planar orientation during casting. The higher crazing rate of the higher molecular-weight polymer was attributed to more residual stress in the higher molecular-weight film consequent to incomplete annealing.

In an attempt to reconcile the seemingly contradictory results of Hopfenberg and Rudd, the effects of molecular weight and subtle orientation changes on solvent crazing of glassy polystyrene films were studied independently.

Two narrow molecular-weight distribution samples of polystyrene of molecular weight 116,000 and 537,000 were dissolved in benzene and cast on a mercury surface producing uniform films of 6.0 ± 0.1 mil thickness. Alternatively, these polymer solutions were cast on smooth glass plates using a hand draw-down with a uniform gap casting bar. In both cases, the films were dried in air overnight and then placed in a vacuum oven at 40 C for 24 hours. After annealing these films for 15 minutes at 105 C, the samples were immersed in n-pentane overlaying mercury in a crystallizing dish maintained at constant temperature. Crazing resulted from the pentane immersion over the range of temperatures studied. After crazing for a specified time, the samples were removed and placed in isooctane at 0 C for 15 min to quench the crazing and exchange residual crazing solvent. Progress of crazing was studied by microtoming cross sections of the crazed samples and measuring the core thickness as a function of time with a filar micrometer eyepiece mounted on a light microscope.

A comparison of the rates of crazing is presented in Figure 10; these data refer to crazing at 35 C. Clearly, the craze front is moving to the film midplane at a constant velocity in both cases and, therefore, relaxations at the craze front are controlling the crazing kinetics.

The rate of crazing is faster for the lower molecular-weight sample. These data are consistent with the early data of Rudd. Presumably, the mercury cast films are essentially stress free consequent to annealing; hence, these data relate uniquely to the effect of molecular weight on the rate of crazing. It is quite reasonable that the low molecular-weight sample relaxes more rapidly and so crazes more rapidly than the high molecular-weight specimen.

The temperature dependence of the crazing process is described in the Arrhenius plots of Figure 11. Crazing in both samples is controlled by extremely highly activated relaxations; the slopes of both lines reflect activation energies in excess of 60 kcal/g-mole.

Films of the same polystyrene samples were cast on glass in an attempt to follow the earlier procedure of Hopfenberg[20] which yielded results suggesting that higher

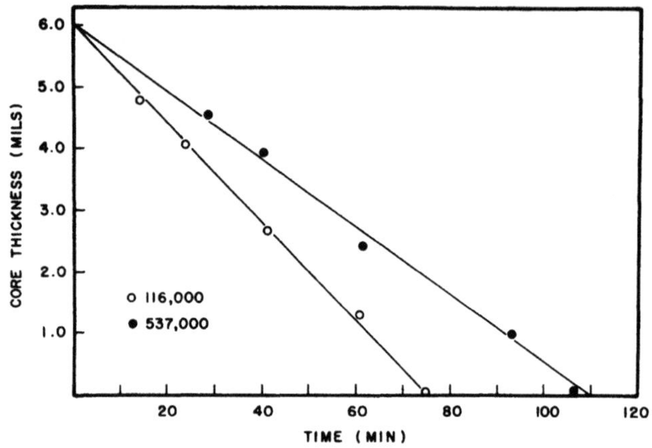

Figure 10. Comparison of crazing rate at 35 C of 116,000 and 537,000 molecular weight polystyrene film cast on mercury.

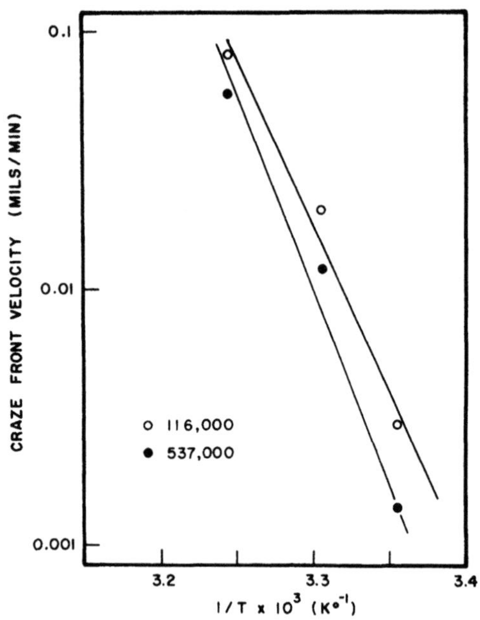

Figure 11. Arrhenius plot of craze-front velocity as a function of reciprocal temperature.

molecular-weight polystyrene crazed more rapidly than low molecular-weight polystyrene. The crazing kinetics clearly indicated that the craze boundary moved at a constant velocity toward the film midplane for these glass-cast films.

The data for n-pentane crazing of glass-cast polystyrene are summarized and compared with the data for mercury cast films in Table 2.

TABLE 2. COMPARISON OF CRAZE-FRONT VELOCITY OF MERCURY- AND GLASS-CAST POLYSTYRENE FILMS AT 35 C

	Craze-front Velocity, mils/min	
	mw = 116,000	mw = 537,000
Glass-Cast	0.0860	0.1020
Mercury-Cast	0.0795	0.0525

The data are rather striking in that the higher molecular-weight film did indeed craze more rapidly than did the low molecular-weight film. These results strongly reinforce the early hypothesis that this apparent anomaly occurs from the residual orientation in the film which results from the rapid draw-down of the casting bar and the restraints of the solid substrate. Presumably, this casting-induced orientation is not completely removed by the rather mild annealing conditions of 105 C for 15 min. Presumably, annealing these samples at a temperature markedly above Tg (ca. 160 C) for extended periods (ca. 2000 min) should remove the residual orientation in the samples which apparently obscure the effect of molecular weight per se. These studies are continuing in our laboratory in a systematic study of the effect of molecular weight, molecular weight distribution, and orientation on the sorption and crazing behavior of polystyrene.

ACKNOWLEDGMENT

I wish to thank Dr. V. Stannett for his many comments and suggestions which aided materially in the preparation of this manuscript.

REFERENCES

(1) J. Crank and G. S. Park (ed), *Diffusion in Polymers*, Chapters 3 and 5, Academic Press, New York, N. Y. (1968).

(2) A. S. Michaels, Chapter "Ultrafiltration" in "Progress in Separation and Purification", Ed. E. S. Perry, Interscience, New York, p. 297 (1968).

(3) R. M. Barrer and G. Skirrow, *J. Polymer Sci.,* **3**, 549 (1948).

(4) A. Aitken and R. M. Barrer, *Trans. Faraday Soc.,* **51**, 116 (1955).

(5) S. Prager and F. A. Long, *J. Am. Chem. Soc.,* **73**, 4072 (1951).

(6) R. J. Kokes and F. A. Long, *J. Am. Chem. Soc.,* **75**, 6142 (1953).

(7) H. Fujita, A. Kishimoto, and K. Matsumoto, *Trans. Faraday Soc.,* **56**, 424 (1960).

(8) G. S. Park, *Trans. Faraday Soc.,* **48**, 11 (1951).

(9) L. Mandelkern and F. A. Long, *J. Polymer Sci.,* **6**, 457 (1951).

(10) M. J. Hayes and G. S. Park, *Trans. Faraday Soc.,* **51**, 1134 (1955).

(11) T. Alfrey, E. F. Gurnee, and W. O. Lloyd, *J. Polymer Sci.,* **C12**, 249 (1966).

(12) T. Alfrey, *Chemical and Engineering News,* **43**, No. 41, 64 (1965).

(13) A. S. Michaels, H. J. Bixler, and H. B. Hopfenberg, *J. Appl. Poly. Sci.,* **12**, 991 (1968).

(14) H. B. Hopfenberg, R. H. Holley, V. Stannett, *Polymer Engineering and Science,* 9, 242 (1969).

(15) T. W. Kwei and H. M. Zupko, *J. Polymer Sci.* (In press).

(16) H. L. Frisch, T. T. Wang, and T. K. Kwei, *J. Polymer Sci.* (In press).

(17) G. A. Bernier and R. P. Kambour, *Macromolecules,* **1**, 393 (1968).

(18) J. F. Rudd, *J. Polymer Sci.,* **B1**, 1 (1963).

(19) R. J. Cresa, "Block and Graft Copolymers", p. 97, Butterworths, London (1962).

(20) H. B. Hopfenberg, Ph.D. Thesis, M.I.T., Dept. of Chemical Engineering, Cambridge, Mass. (1964).

(21) G. King, *Trans. Faraday. Soc.,* **41**, 325 (1945).

(22) J. Crank, *J. Polymer Sci.,* **11**, 151 (1953).

(23) J. Bray and H. B. Hopfenberg, *J. Polymer Sci.,* Part B (in press).

(24) H. B. Hopfenberg and H. L. Frisch, *J. Polymer Sci.,* Part B, **7**, 405 (1969)

APPLICATION AND THEORY OF MEMBRANE PROCESSES FOR BIOLOGICAL AND OTHER MACROMOLECULAR SOLUTIONS

Richard P. deFilippi and Robert L. Goldsmith
ABCOR, INC.
Cambridge, Massachusetts

BACKGROUND

The application of membrane processes to industrial separations is becoming a reality. For operations requiring concentration and purification of aqueous streams ranging from fermentation broths containing products such as enzymes and pharmaceuticals, to brackish and polluted waters, small scale commercial membrane units are presently in operation. Many of these involve the process of membrane ultrafiltration, which process has demonstrated a capability for treating a variety of industrial solutions.

With the extension of membrane ultrafiltration to a large number of systems, a significant body of information on process behavior has been generated. For example, in our laboratories we have studied the application of ultrafiltration to over 25 industrial cases. With that accumulated information, it is becoming possible to establish a process model which has significant value in data correlation for process and equipment-design purposes. The description of such a model and experimental results supporting it are discussed in this paper.

It should be recognized that a full understanding of membrane ultrafiltration has yet to be achieved. Thus, design for each application requires a basic body of experimental data. Experience indicates that no single equipment design or membrane configuration will be optimum for all applications. Indeed, the purpose of modeling based on engineering fundamentals is to allow choice of an optimum design based on the minimum necessary quantity of data.

Membrane ultrafiltration is a process which is capable of separating components of a solution largely on the basis of molecular size. Separation can be effected between solvent and solute, or between different solutes in a multicomponent solution. The separating agent is a thin membrane which can be viewed as a molecular screen, characterized by a pore size which will allow transport of solvent and lower molecular-weight solutes, and hinder passage of higher molecular-weight solutes. Although transport is pressure-driven, many cases exhibit the apparent anomaly that flow-rate is independent of the magnitude of the operating pressure. Ultrafiltration should be distinguished from a similar process, reverse osmosis, which is capable of separating water from small molecules such as dissolved salts. In reverse osmosis, the solution osmotic pressure is significant (on the order of 350 psia for sea water, for example), and the applied pressure must exceed the osmotic pressure in order to achieve transport of water through the membrane, away from the high concentration solution. In ultrafiltration, the macromolecules which are screened, or retained, account for only a small osmotic pressure difference across the membrane. As a result, ultrafiltration operating pressures can be in the range of 10 psi, in contrast to several hundred psi required for reverse osmosis.

Recent developments in ultrafiltration membrane technology have been important factors in bringing the process into the range of economic feasibility. Principal among these are the development of high flux, asymmetric membranes of cellulose acetate as well as other materials[(1,2)], and more recently, small hollow fiber membranes[(3,4,5,6)]. Full discussion of commercially available membranes is not within the scope of this paper. It is important to indicate, however, that these developments have led to "tailored" membranes, the retentive properties of which can be controlled to give the desired separation in specific applications.

APPLICATIONS

Industrial applications of ultrafiltration may be categorized broadly within the following three groups:

Biological Products. Products from biological processes (fermentations and cultures), such as pharmaceuticals and enzymes, can be recovered from complex aqueous solutions. As a wide range of solutes are generally present, ultrafiltration can be used to retain the product, which is often of moderate to high molecular weight, while transmitting water and low molecular weight solutes through the membrane. This produces a concentrate which can be dried to obtain the final purified solid product.

Food and Beverage Products. Products from food processing operations such as those in the dairy, potato, and corn industries, may be treated by ultrafiltration to yield potentially high-value byproducts. In particular, whey resulting from cheese production is rich in both proteins and sugars which can be recovered by a combination of ultrafiltration and reverse osmosis.

Waste Treatment. Ultrafiltration can be used to reduce the pollution contribution of organic solutes in wastes from production of industrial chemicals, pulp and paper, and food products.

In Tables 1 and 2, pilot-plant data are presented for two of the applications described above: industrial enzymes, and cottage-cheese whey. In Table 1, two separate examples of enzyme ultrafiltration are shown: one involves a protease solution, and the other an amylase/protease mixture. For the amylase/protease mixture, using an Abcor HFA-200 membrane at a pressure of 15 psi, a steady-state flux of 11 gal/day/ft^2 was observed with high retention of the product enzyme (96 percent or greater). For the protease, employing the same membrane at 20 psi, a steady-state flux of 7.8 gal/day/ft^2 was observed, with an enzyme retention of 99 percent. In this case, the retention of total solids was also measured, and was found to vary in the range of 15 to 30 percent, depending on feed concentration. Corresponding to this relatively low solids retention, the majority of solutes passed through the membrane which quantitatively retained the enzyme. This resulted in an enzyme product of increased specific enzyme activity/unit weight of product, relative to that produced by techniques which cause retention of all solids, such as evaporation. Also, passage of most of the solids into the permeate allows concentration of enzymes without a significant increase in the viscosity of the solution. As will be shown, low viscosity greatly facilitates process operation.

In Table 2, the results of pilot-scale studies on the ultrafiltration of cottage-cheese whey are summarized. In these studies, the objective was to provide an initial fractionation which would concentrate and purify whey proteins, by removing most other solutes, including lactose, amino acids, lactic acid, and salts. The permeate from the first membrane stage containing these materials was then passed to a second stage containing a "tighter" membrane with a very low molecular-weight cutoff, designed to retain lactose and other organics contributing to the high chemical-oxygen-demand (C.O.D.) level in the raw whey. The permeate from the final stage can be reused within the dairy as fresh water. Whey proteins and lactose recovered from the concentrates from the two stages can be recovered as valuable process products.

TABLE 1. ULTRAFILTRATION OF INDUSTRIAL ENZYME SOLUTIONS

	Amylase/Protease (1% Total Solids)	Protease (10% Total Solids)
Membrane	HFA-200	HFA-200
Pressure, psi	15	20
Flux, gfd	11	7.8
Enzyme retention, %	97 (amylase) 96 (protease)	99
Total Solids Retention, %	--	15-30

TABLE 2. ULTRAFILTRATION OF COTTAGE-CHEESE WHEY

	Feed	First-Stage Permeate	Second-Stage Permeate
Total Solids, %	7.8	6.8	1.8
C.O.D., ppm	65,600	54,100	800
Protein and Amino Acid Nitrogen, %	0.6	0.15	0.002
Lactose, %	3.9	3.5	0.05
Lactic Acid, %	0.52	0.52	0.11

The results show that the majority of solids were transmitted through the first stage membrane, while protein retention was high. Almost all of the lactose, lactic acid, and amino acids were transmitted through the first-stage membrane. The second-stage membrane was effective in retaining the major portion of solids, including lactose in high yield. While lactic acid was partly transmitted through the membrane in the second stage, the C.O.D. was nevertheless drastically reduced. Thus, employing two membrane types tailored for this separation, it is possible to obtain two products of value from cheese whey, protein and lactose, while effecting nearly quantitative removal of organic pollutants.

PROCESS ANALYSIS AND EXPERIMENTAL STUDY

To gain a basic understanding of flux-controlling variables in ultrafiltration, it is necessary to analyze the fundamental mass-transfer operations occurring. This must include consideration of transport in fluid phases adjacent to the membrane, as well as transport within the membrane itself. A basic analysis describing transport behavior in the fluid phase in membrane processes such as ultrafiltration was originally developed by Brian[7] for reverse osmosis and later extended to ultrafiltration by Michaels[8].

Figure 1 depicts the concentration profile near the membrane surface at steady-state conditions. If the membrane exhibits high retention for a given solute at an average upstream concentration, C_b, there will be an increase in concentration of solute as the distance to the

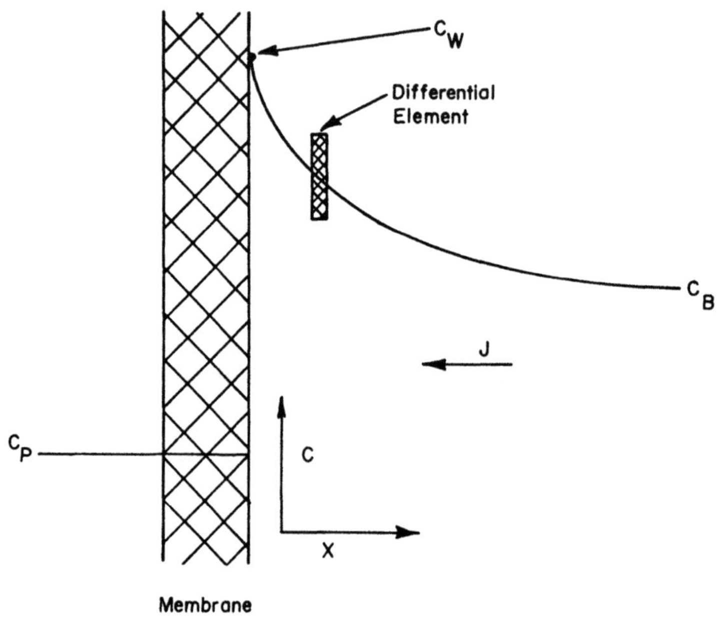

Figure 1. Concentration polarization, schematic.

membrane surface decreases, reaching a value of C_W at the surface. Solute brought to the membrane surface by the permeation flow of solution is balanced by diffusion back into the bulk solution under the influence of the concentration gradient away from the membrane surface and by solute transport through the membrane. Neglecting axial concentration gradients, a material balance in the region near the upstream surface of the membrane yields the equation:

$$J = k \ln \left[\frac{C_W - C_p}{C_b - C_p} \right] \tag{1}$$

Here, J is the solution flux through the membrane; C_p, the permeate concentration; and k, a mass transfer coefficient. The latter coefficient is used assuming it is equal to D/δ, where D is the solute diffusivity, and δ is the concentration boundary-layer thickness.

The practical effect of the concentration polarization depicted in Figure 1 on ultrafiltration flux is relatively small provided that C_W is below the solute solubility limit. While the elevated concentration level at the membrane surface may cause an increased permeate solute concentration, which is undesirable, there would generally be a negligible effect on solution flux through the membrane, as the increased concentration at the membrane surface, C_W, would bring about only a slight increase in osmotic pressure gradient across the membrane. This effect is very small in ultrafiltration because of the high molecular weight of the retained solute, and thus its very low osmotic pressure.

In spite of this, however, two important experimental facts have been noted:

(1) Membrane fluxes for most macromolecule solutions are significantly lower than those measured for water at the same operating conditions. In the absence of osmotic pressure effects, one would expect these fluxes to be equivalent.

(2) With macromolecule solutions, membrane fluxes have been found to be independent of pressure; in contrast, fluxes with pure water generally increase linearly with pressure.

Thus, it is necessary to account for these observations by an additional mechanism.

In Figure 2, the concentration profile near the membrane surface is depicted, as originally presented by Michaels(8). In the case of macromolecular solutions, the concentration at the membrane surface can build up to the point where formation of a precipitate, or gel layer, can occur. For ultrafiltration in general, high concentration polarization ratios, C_W/C_b, would be expected, since a high concentration gradient away from the surface is required to effect back diffusion into the solution, as high molecular-weight solutes have very low diffusion coefficients. The presence of the gel layer influences the permeation flux because it introduces a hydraulic resistance often significantly greater than that of the membrane. That is, transport through the membrane is now controlled by the permeation properties of this gel layer.

Under conditions of gel formation, then, the polarization term in Equation (1), $\ln [(C_W - C_p)/(C_b - C_p)]$, is fixed at a given bulk concentration, because C_W is fixed by solubility

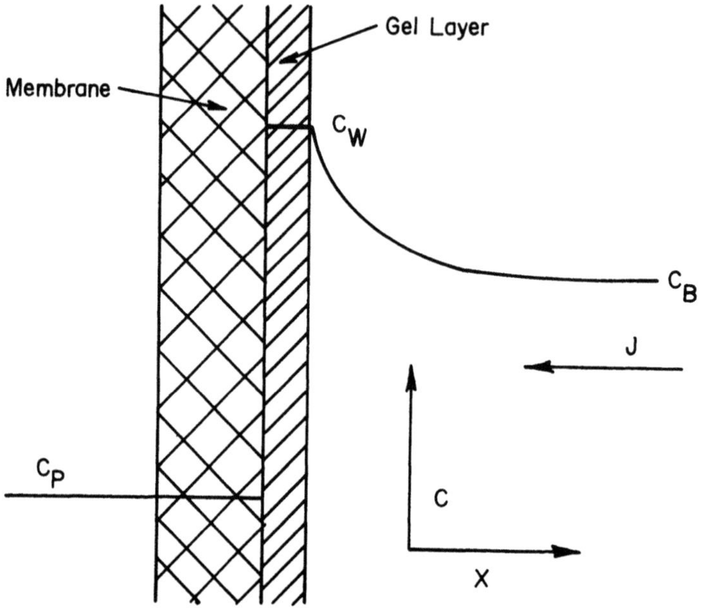

Figure 2. Gel formation, schematic.

considerations. If retention of solute is high, C_p is small and changes in retention will have a small effect. Thus, the solution flux, J, is directly proportional to the mass transfer coefficient, k, and decreases logarithmically with feed concentration. If these assumptions are correct, those fluid properties and system parameters which influence the mass transfer coefficient are the variables which will influence membrane flux, J.

This mechanism also explains the lack of pressure dependence of flux in the ultrafiltration of macromolecular solutions. When pressure is increased, a transient increase in solution flux is indeed observed. However, at a new steady state, the transport of solute to the membrane surface due to permeation of solvent must still be balanced by the diffusion of accumulated solute at the surface back into the bulk solution. Increased pressure does not increase this back diffusion effect; consequently there must be a transient rate of accumulation of solute at the membrane surface which ceases only at the point where the forward and reverse transport of solute at the surface again become equivalent. This must occur when the permeation resistance of the gel layer has increased just to the extent necessary to counterbalance the higher solution flux due to the pressure increase. Thus, the new steady-state membrane flux becomes equivalent to that which prevailed prior to the pressure increase.

Under conditions of fully developed turbulent flow, the relationship between mass transfer coefficient, system properties, and flow conditions has been given as[9]:

$$\frac{kd}{D} = A_1 \, (Sc)^{1/3} \, (Re)^{4/5} \qquad (2)$$

where d is the hydraulic diameter of the flow channel; Sc is the Schmidt number ($\mu/\rho D$); Re is the Reynolds number ($du\rho/\mu$); and A_1 is a constant.

In fully developed laminar flow, under conditions where the concentration boundary layer is developing, which can be the entire channel length for fluids with high Schmidt numbers, the mass transfer coefficient is given by[10]:

$$\frac{kd}{D} = A_2 \left(Re \cdot Sc \cdot \frac{d}{L} \right)^{1/3} \tag{3}$$

where L is the channel length, and A_2 is a constant. In the so called "far downstream" region, where the concentration profile is fully developed, the following relationship is applicable in laminar flow[10]:

$$\frac{kd}{D} = A_3 \tag{4}$$

Brian[7] has presented the results of numerical solutions to the laminar flow concentration polarization problem, accounting for axial concentration differences down the flow channel. Dresner[10] has developed approximate analytical solutions to the same problem. These solutions have been compared with Equation (1) with k defined by Equation (3). The latter predicts results equivalent to the Brian and Dresner solutions to within ± 15 percent.

In the special case in laminar flow where the velocity and concentration profiles both are developing, the following relationship is applicable[11]:

$$\frac{kd}{D} = A_4 (Sc)^{1/3} \left(Re \cdot \frac{d}{L} \right)^{1/2} \tag{5}$$

All of the above relationships are based on constant system properties, such as diffusivity and viscosity. Since concentrations of macromolecular solutes vary significantly from bulk solution to the membrane surface, there are variations in concentration-related system properties. The most significant is probably viscosity, which may vary as much as ten-fold between the highly concentrated solution at the membrane and the more dilute concentration in the bulk stream. At present, analytical solutions to the case of variable viscosity have not been reported. However, this has been examined experimentally for the analogous heat transfer situation, where temperature effects on viscosity are significant[12]. Applying the viscosity relationship found in heat transfer to the mass transfer case, an additional term $(\mu_w/\mu_b)^n$ should be factored into Equations (2) through (5), where μ_w represents the viscosity of fluid at the wall, and μ_b the viscosity of fluid in the bulk stream. In heat transfer, the exponent n has a value of 0.14. Because of this relatively small value, assuming that the analogy with heat transfer is valid, a viscosity adjustment should have a relatively small effect on the mass transfer coefficient in most cases.

This model, then, provides relationships between membrane solution flux, under conditions of gel formation, to properties of the fluid and system hydrodynamics using established mass transfer relationships. These relationships can serve as the basis for data correlation and, more important, for interpolation in design optimization. To determine the validity of this approach, an experimental program was carried out with model solutions of known viscosity and diffusivity over a wide range of Reynolds numbers, and in two different flow-channel geometries. Membranes of known transport properties were used.

Test runs were carried out in a unit similar to that shown in Figure 3. In turbulent flow, a 1-in. diameter tubular membrane was used. In laminar flow, a thin-channel membrane unit of rectangular cross-section was used, with a channel height of 60 mils, a length of 10 in. and a width of 5 in. All experiments were performed under conditions of differential operation; that is, axial concentration changes were negligible. Both permeate and retentate were recycled to the feed solution reservoir, so that steady-state operation at a fixed-feed concentration was achieved.

The solute employed was Carbowax 20M®, a polyethylene glycol with a mean molecular weight of 15,500. It was subsequently learned that the molecular weight distribution of this material is bimodal; this may have some influence on solute retention data, but should not be important in affecting flux. Retentions were sufficiently high so that almost all solute was retained by the membrane. Consequently, properties of the gel formed at the upstream face of the membrane should not have been influenced by variations in the average molecular weight of the retained material.

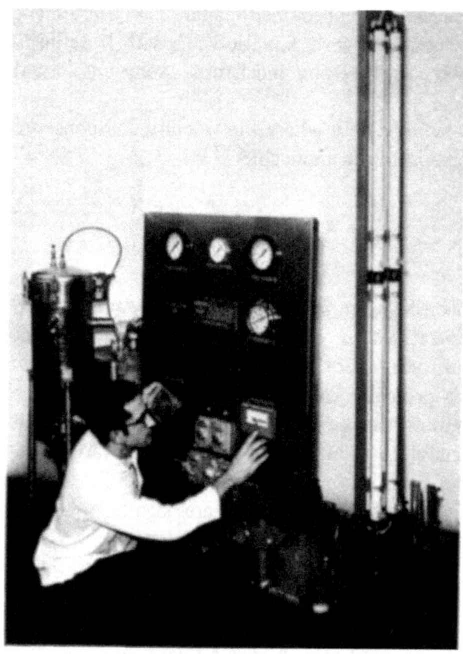

Figure 3. Front view of UF-18 ultrafiltration pilot plant.

(6 sq ft of tubular HFA membrane installed)

® Registered trademark of Union Carbide Company.

The influence of Reynolds number on flux is shown in Figures 4 and 5 for turbulent and laminar flow, respectively. In turbulent flow, flux data covering a Reynolds number range from about 6,000 to 50,000 agreed very well with a power-law relationship, with a Reynolds number exponent of 0.87. This is somewhat higher than the 0.8 power predicted by Equation (2). It should be pointed out, however, that studies of mass transfer in solutions with high Schmidt numbers(13) indicate that an exponent of about 0.9 may be more accurate under these conditions. If the value of 0.023 is assigned to the constant A_1 in Equation (2)(9), using the measured bulk solution viscosity of 1.3 centipoises, a diffusivity of 0.3 x 10^{-6} cm^2/sec was calculated. This compares favorably with the reported value(14) of 0.5 x 10^{-6} cm^2/sec.

In laminar flow, as shown in Figure 5, an excellent fit of the flux-Reynolds number data can be obtained using an exponent of 0.50. From the relationship:

$$\frac{L^*}{d} = 0.05\ Re \tag{6}$$

the channel length L^* required to attain fully developed laminar flow can be calculated, assuming a flat velocity profile at the channel inlet. This calculation shows that at the higher Reynolds numbers, about 30 percent of the channel length was needed for velocity boundary-layer development, during which Equation (5) would be applicable in describing mass transfer. While the velocity boundary layer development length is less at the lower Reynolds numbers, apparently the mass transfer relationship holds throughout the Reynolds number range studied. Using a value of 0.664 for the constant A_4 in Equation (5). A diffusivity of 0.7 x 10^{-6} cm^2/sec has been calculated, which is in reasonably good agreement with the reported value of 0.5 x 10^{-6} cm^2/sec. Other investigators(16) have reported ultrafiltration data for solutions in fully developed laminar flow. These show reasonably good agreement with the behavior predicted by Equation (3).

In previously reported data, both for this test solution(17) and other systems(16,17), the relationship given in Equation (1) between flux and solution concentration at constant Reynolds number has been demonstrated to be applicable over wide ranges of concentrations. These data provide further support for the applicability of this mass transfer model.

Considering the membrane as a molecular screen, one would expect that the permeate concentration of solute should be a function of solute concentration at the membrane surface. Under conditions of gel formation, where the surface concentration would be expected to remain constant, and independent of bulk-stream concentration, it would be anticipated that the permeate concentration would similarly remain unchanged over a range of bulk-stream concentrations.

To test this hypothesis, permeate concentrations were measured as a function of feed concentration in the polyethylene glycol system in both turbulent and laminar flow. Results are given in Figures 6 and 7. It is evident that permeate concentration does not remain constant as the feed concentration is increased. Indeed, it appears that permeate concentration increases approximately in proportion to the feed concentration. This is inconsistent with the assumption of unchanging concentration at the membrane surface. However, this assumption appears to be borne out by flux-concentration behavior. At present, these results are unexplained. It should be noted, however, that various other systems exhibit the same behavior, i.e. Dextrans, while protein solutions show constant permeate concentration, with increasing feed concentration, consistent with predictions(17).

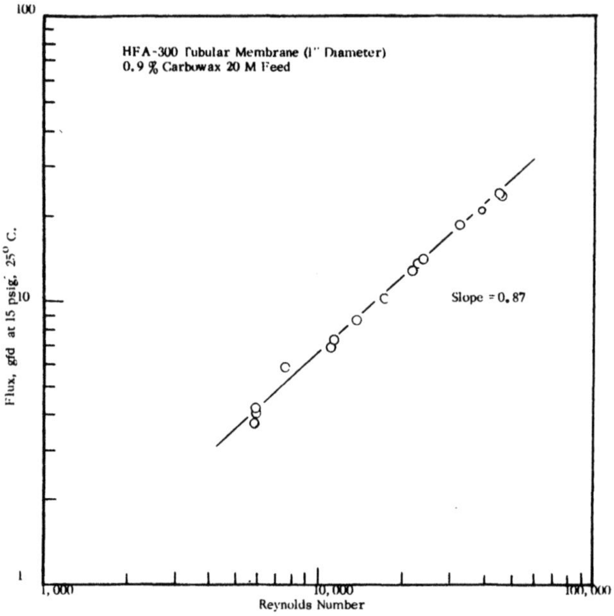

Figure 4. Ultrafiltration flux in turbulent flow.

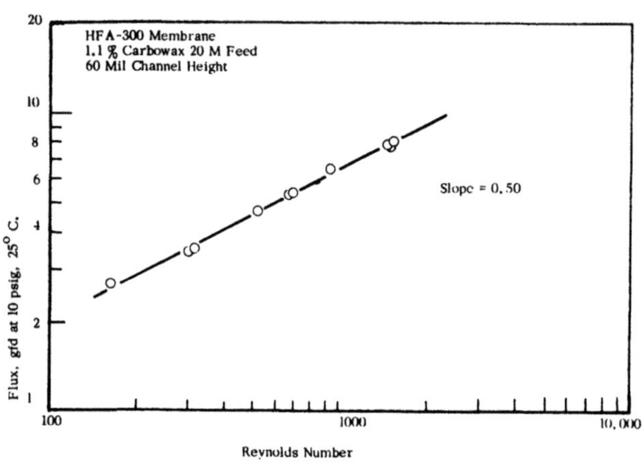

Figure 5. Ultrafiltration flux in laminar flow.

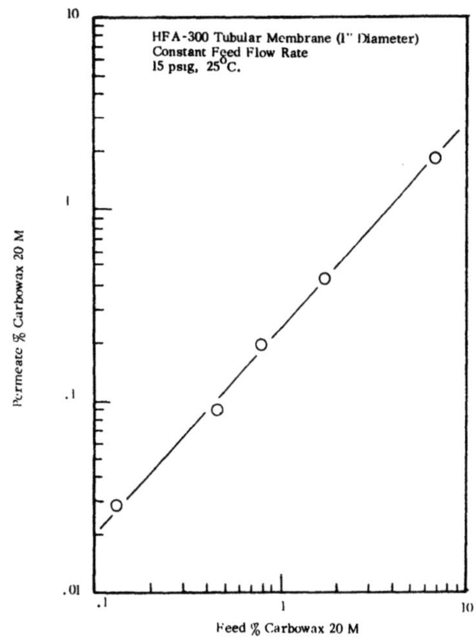

Figure 6. Solute retention in turbulent flow ultrafiltration.

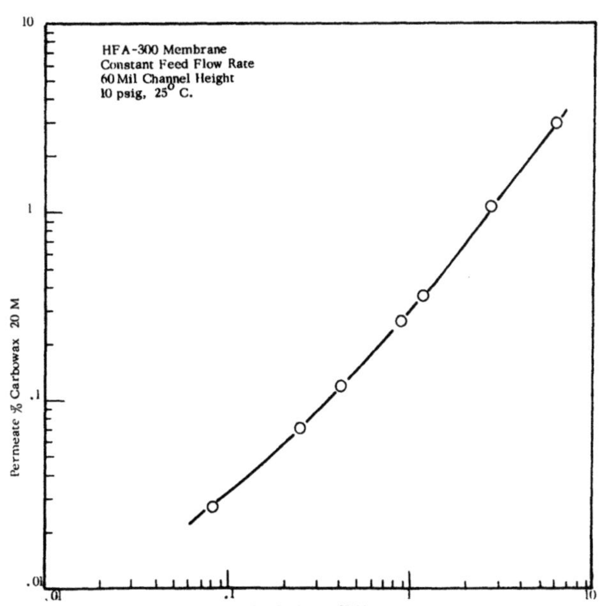

Figure 7. Solute retention in laminar flow ultrafiltration.

APPLICATIONS TO PROCESS DESIGN

Accepting the validity of this mass-transfer model for ultrafiltration, it is possible to develop a rational basis for process design. As will become apparent, no single design of ultrafiltration equipment and processes will be optimal for all applications. However, given fluid properties, it is possible for the engineer to make a reasoned choice of independent variables at his disposal: flow-channel dimensions, such as length and diameter (or height); and fluid-flow rates, which may be set independently by use of feed recirculation. High-flux membranes are available both in tubular and flat-film forms; thus, ultrafiltration processes will be operated in flow regimes where membrane resistance is unimportant, and fluid-phase mass transfer controls flux. While this does not represent the state of the art in capillary-membrane systems, it is probable that high-flux capillary membranes will be available in the future. Thus, it also is worthwhile considering their use in high-flux ultrafiltration.

To minimize capital investment for membranes, as well as the operating cost of membrane replacement, it is obviously desirable to operate at as high a membrane flux as possible. Under conditions of gel formation, high fluxes are achieved at high Reynolds number which, in turn, are achieved at the expenditure of pumping power. The mass transfer relationships described previously provide the basis for computing flux as a function of Reynolds number, if viscosity, diffusivity, channel diameter, and channel length are known. When these properties are fixed, one may also calculate power requirements per unit volume of permeate production from straightforward hydrodynamic equations. An example of a computation of this kind is summarized in Figure 8.

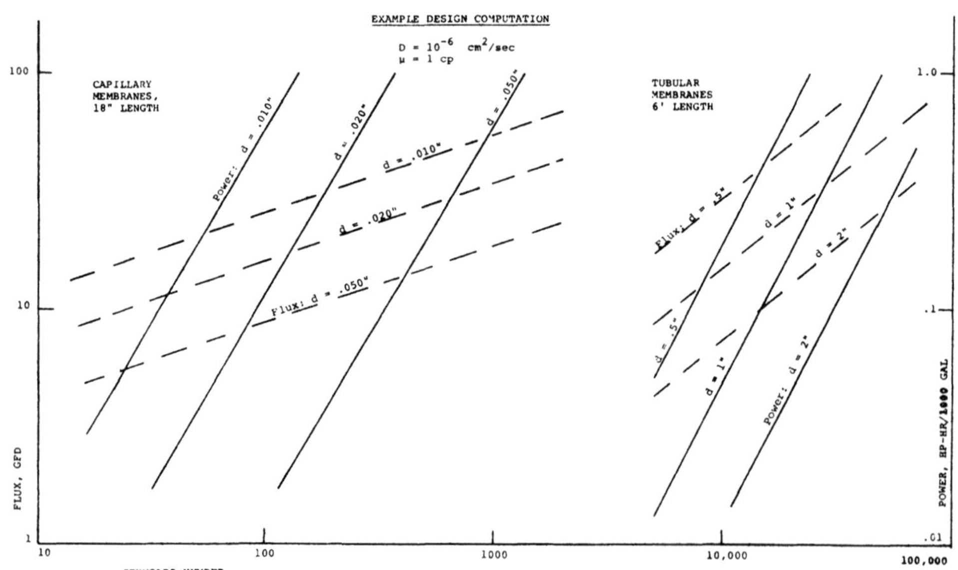

Figure 8. Ultrafiltration flux and power requirements.

In this example, both flux and power consumption have been computed as a function of Reynolds number in both laminar- and turbulent-flow ultrafiltration for three different tube diameters in each case. In turbulent flow, tubular membranes 6 ft in length were chosen with diameters ranging from 0.5 to 2.0 in. In laminar flow, high-flux capillary membranes 18 in. long having diameters ranging from 0.01 to 0.05 in. were chosen. Computation for the latter system would also be applicable to thin-channel, flat-film membrane systems as well. Dashed lines in Figure 8 refer to flux, given on the left-hand ordinate in gal/day/ft^2. Solid lines refer to power, given on the right-hand ordinate as hp-hr/Mgal of permeate. A diffusivity 10^{-6} cm^2/sec and a viscosity of 1 centipoise were assumed.

For the conditions specified in this example, first it is apparent that power requirements are very low, and high Reynolds numbers result in optimum performance. The maximum worthwhile Reynolds number would be determined by the flux of the membrane alone; once the solution flux becomes equivalent to the pure-water flux of the membrane, further increases in Reynolds number would give no additional benefit, since this represents the point at which the gel layer is eliminated at the membrane surface. Second, small-diameter channels provide the highest flux for a given flow regime.

It is instructive to consider qualitatively the trends that may occur as system properties are varied from those used in this sample calculation. For a lower diffusivity, all flux lines will be lowered by the same factor. Because of the higher exponent on the flux-Reynolds number functionality in turbulent flow, it would be possible to achieve increased fluxes more readily in turbulent flow than in laminar flow by simply increasing Reynolds number. Thus, the increase in power costs necessary to achieve high fluxes would be less in turbulent flow than in laminar flow, which would indicate that the turbulent-flow case might be the more favorable alternative under these conditions.

As viscosity is increased, Equation (3) shows that flux in laminar flow is unaffected. According to Equation (2), however, flux will decrease with increasing viscosity in turbulent flow. Since viscosity has a very large effect on power consumption, it would appear that turbulent flow would become unfavorable for achieving high fluxes in a region of high viscosity, thus favoring the laminar-flow approach.

SUMMARY

An analysis, supported by experimental evidence, shows that the flux behavior of ultrafiltration systems can be modeled using known chemical engineering mass-transfer relationships, although prediction of solute retention is hazardous in the absence of data. Given experimental information on flux as a function of Reynolds number and solute concentration, relationships can be obtained which will provide the basis for a design optimization. This has utility for projecting preliminary economics of a process and for guiding further experimentation prior to commitment to large-scale equipment. Design computations for a variety of systems show that no single equipment and process design is optimum for all ultrafiltration applications. Rather, it is necessary to obtain pilot-plant information on flux and retention for the specific separation of interest to provide the basis for determining the optimum process design.

REFERENCES

(1) S. Loeb, *Desalination by Reverse Osmosis,* Edited by U. Merten, MIT Press, p 55 (1966).

(2) A. S. Michaels, *Advances in Separations and Purifications,* Edited by E. S. Perry, John Wiley & Sons (1968).

(3) R. J. Mattson, and V. J. Tomsic, *Chem. Eng. Prog.,* **65** (1), p 62 (1969).

(4) H. I. Mahon, E. A. McLain, W. E. Skiens, B. J. Green, T. E. Davis, Chem. Eng. Prog., Symp. Ser., **69** (91), pp 48-51 (1969).

(5) J. D. Bashaw, and T. A. Orofino, "Hollow Fibers for Reverse Osmosis", presented at the Second Symposium on Reverse Osmosis, Miami, Fla. (1969).

(6) R. P. deFilippi, R. L. Goldsmith, and R. S. Timmins, "Development of a Water Desalination System Based on Hollow Fine Fibers", ibid.

(7) P.L.T. Brian, *Desalination by Reverse Osmosis,* Edited by U. Merten, MIT Press, p 161 (1966).

(8) A. S. Michaels, *Chem. Eng. Prog.,* **64**, No. (12), p 31 (1968).

(9) R. E. Treybal, *Mass Transfer Operations,* McGraw Hill (1955).

(10) T. R. Sherwood, P.L.T. Brian, R. E. Fisher, and L. Dresner, *Ind. & Eng. Chem. Fundamentals,* **4**, p 113 (1965); L. Dresner, Oak Ridge Nat'l. Lab. Report 3621 (May, 1964). (Cited in Ref. 7)

(11) H. Grober, S. Erk, and V. Grigull, *Fundamentals of Heat Transfer,* p 233, McGraw Hill (1961).

(12) E. N. Sieder, and G. E. Tate, *Ind. & Eng. Chem.,* **28**, p 1429 (1936).

(13) P. Harriott, and R. M. Hamilton, *Chem. Eng. Sci.,* **20**, p 1073 (1965).

(14) J. Brandup, and E. H. Immergut, *Polymer Handbook,* Interscience (1967).

(15) W. M. Kays, *Convective Heat and Mass Transfer,* McGraw Hill, p 63 (1966).

(16) H. J. Bixler, R. W. Hausslein, and L. Nelson, "Separation and Purification of Biological Materials by Ultrafiltration", presented at the 65th National Meeting of the American Institute of Chemical Engineers, Cleveland, Ohio (May 4 - 7, 1969).

(17) R. L. Goldsmith, "Macromolecular Ultrafiltration with Microporous Membranes", presented at the 158th National ACS Meeting, New York (Sept. 1969).

SOLUTE POLARIZATION AND CAKE FORMATION IN MEMBRANE ULTRAFILTRATION: CAUSES, CONSEQUENCES, AND CONTROL TECHNIQUES

William F. Blatt, Arun Dravid, Alan S. Michaels, and Lita Nelsen
Amicon Corporation, Lexington, Massachusetts

INTRODUCTION

In the past 5 years, membrane ultrafiltration has gained increasing prominence as a simple and convenient process for concentrating, purifying, and fractionating solutions of moderate-to-high molecular weight solutes and colloids, and for purifying water and other solvents containing such solutes. The emergence of this new molecular separation technique for both laboratory and industrial applications is almost entirely attributable to the development of a family of uniquely structured polymeric membranes which display extraordinarily high hydraulic permeabilities coupled with the capacity to retain even quite small solute molecules.

Efforts to make efficient use, in practical separations, of the rather remarkable properties of these highly permeable molecular filters have in no small measure been frustrated by a phenomenon which is ascribable to their "high-permeability". This is the phenomenon of "concentration-polarization" – an effect long recognized by electrochemists but, until recently, little appreciated by physical chemists and chemical engineers. In the simplest terms, concentration-polarization is the accumulation, at the upstream surface of an ultrafiltration membrane, of solute molecules which are rejected or retained by the membrane in the course of ultrafiltration. At the outset, it suffices to say that concentration-polarization always operates to reduce the efficiency and/or rate of an ultrafiltration process. It is toward this phenomenon – its causes, effects, and techniques for its control – that this discussion is directed.

An understanding and appreciation of concentration-polarization, as well as its quantitative analysis, requires an understanding of the basic transport processes which take place within ultrafiltration membranes. Membranes which display the capacity to retain rather small solute molecules (i.e., those of molecular weight under ca 500, or of molecular dimensions under about 10 Å) generally transmit solute and solvent molecules via molecular diffusion within the membrane matrix. When a solution is confined under pressure on one side of such a membrane, the solvent and solute transport rates across the membrane can be approximated by the relationships

$$J_1 \simeq \frac{\bar{P}_1}{t_m} (\Delta P - \Delta\Pi) \tag{1}$$

$$J_2 \simeq \frac{\bar{P}_2}{t_m} (C_B - C_f) \tag{2}$$

where J_1 and J_2 are the solvent and solute fluxes across the membrane; ΔP, the hydraulic pressure difference between the upstream solution and the ultrafiltrate; $\Delta\Pi$ the osmotic pressure difference between the two solutions; C_B, the upstream solute concentration; and C_f the solute concentration in the ultrafiltrate. $\bar{P}_1$ and $\bar{P}_2$ are the "specific permeabilities" of the membrane to solvent and solute, respectively; and t_m, the membrane thickness. Mass conservation further requires that the solvent and solute fluxes be related by

$$J_2 = J_1 C_f \tag{3}$$

Simultaneous solution of Equations (1), (2), and (3) yields the important result:

$$1 - \left(\frac{C_f}{C_B}\right) = \sigma = \frac{\bar{P}_1/\bar{P}_2 \ (\Delta P - \Delta\Pi)}{1 + \frac{\bar{P}_1}{\bar{P}_2} \ (\Delta P - \Delta\Pi)} \tag{4}$$

The quantity σ is the "rejection coefficient" of the membrane, i.e., the fraction of the solute present in the upstream solution which is retained by the membrane.

The solvent flux through this "diffusive" type of membrane is thus directly proportional to the "effective pressure difference" across the membrane. On the other hand, the rejection efficiency of the membrane increases hyperbolically with increasing pressure difference. If the solute-permeability of the membrane is sufficiently small relative to its solvent-permeability, the rejection coefficient reaches values approximating unity at quite low pressures.

If solute polarization occurs at the upstream membrane surface, the solute concentration at that surface will be higher than that in the bulk solution. Consequently, according to Equation (1), the solvent flux will be **reduced** because of an increase in $\Delta\Pi$ (at constant ΔP); according to Equations (2) and (3), the solute flux will be increased (since the solute concentration at the upstream membrane surface is elevated) and the rejection efficiency reduced (both by the increase in solute concentration and increase in $\Delta\Pi$). If the ratio of the concentration at the upstream membrane surface to that in the bulk of the solution is represented by M (the "polarization modulus"), the transport relations in the presence of polarization (assuming $C_f << C_B$) are given by

$$J_1 \simeq \frac{\bar{P}_1}{t_m} (\Delta P - M\Delta\Pi) \tag{1a}$$

$$J_2 \simeq \frac{\bar{P}_2}{t_m} (MC_B - C_f) \tag{2a}$$

$$1 - \frac{C_f}{C_B} = \sigma_A = 1 - \frac{M}{1 + \bar{P}_1/\bar{P}_2\ (\Delta P - M\Delta\Pi)} \tag{4a}$$

where σ_A is the "apparent rejection efficiency", computed from the bulk upstream solution concentration. The effect of polarization on rejection is thus particularly great.

Ultrafiltration membranes which display retention for only relatively large solute molecules (those of molecular weight in excess of 500, or of molecular diameters above 10 Å) appear to function as molecular sieves or screens, solvent flowing in viscous flow through micropores in the membrane, and solute molecules being carried convectively with solvent only through the pores large enough to accommodate them. For such membranes, the transport relationships are approximated by

$$J_1 = \frac{\bar{P}_1}{t_m} \Delta P \tag{5}$$

$$J_2 = C_B\,(1 - \sigma)\,J_1 \tag{6}$$

where $\bar{P}_1$ is the hydraulic permeability coefficient of the membrane. The quantity $(1\text{-}\sigma)$ represents the fraction of the solvent flux carried by pores large enough to pass the solute.

The continuity constraint also applies:

$$J_2 = J_1 C_f \tag{3}$$

whereupon:

$$1 - \frac{C_f}{C_B} = \sigma \tag{7}$$

For such membranes, the solvent flux is thus linear in the hydraulic pressure difference, and the rejection coefficient is essentially constant and pressure-independent. The quantity $\Delta\pi$ is absent from the above relations simply because, for high-molecular weight solutes, the osmotic pressure of the retained solute is usually very small compared with the applied hydraulic pressure.

It is frequently found, with microporous ultrafiltration membranes, that if the rejection coefficient for a particular macromolecular solute is less than about 0.9, and greater than about 0.1, the **solvent** flux through the membrane **decreases** with increasing solute concentration in the upstream solution, while the rejection coefficient **increases**. These effects are usually observed only with fairly concentrated solutions – those containing a few percent solute by weight or more. This can be explained by the fact that the viscous drag exerted by solute molecules in the membrane pores increases with solute concentration, thereby reducing flow in those pores passing the solute. This, of course, biases the flow in favor of pores too small to accommodate solute, thus increasing rejection. Obviously, if the membrane is either completely retentive for, or freely permeable to solute, these concentration-dependent effects will be absent.

If polarization of solute occurs upstream of the membrane, the effects can be predicted in terms of the polarization modulus M, as before:

$$J_1 = \frac{\bar{P}_1}{t_m} \Delta P \tag{5}$$

$$J_2 = MC_B\,(1-\sigma)\,J_1 \tag{6a}$$

$$\text{and } 1 - \frac{C_f}{C_B} = \sigma_A = 1 - M\,(1-\sigma) \tag{7a}$$

In this case, one predicts no effect of polarization on flux, and a significant depressing effect upon the apparent rejection coefficient, σ_A. If the membrane shows partial rejection for the solute, however, polarization will **reduce** the flux, and the apparent rejection efficiency – while it will decline with polarization – will decrease somewhat less rapidly than predicted by Equation (7a) because of the viscous drag effect mentioned above. As will be shown in greater detail below, however, phenomena occurring on the upstream surface of the membrane consequent to polarization, which have nothing whatsoever to do with the solvent- and solute-transport processes occurring **within** the membrane, usually so strongly dominate the ultrafiltration process that they mask the effects predicted by Equations (5), (6a), and (7a).

In summary, solute polarization is a natural consequence of ultrafiltration through a membrane displaying solute-retention. The effects of polarization upon "diffusive" and "microporous" membranes – despite the differences in their inherent transport mechanisms – are to reduce flux and/or reduce solute-rejection efficiency. Minimization of polarization is thus a necessary and important objective to achieve maximum performance in ultrafiltrative separation.

MATHEMATICAL MODELING OF CONCENTRATION POLARIZATION

The flux reduction caused by concentration polarization in the ultrafiltration of micromolecular, polymeric, and colloidal solutions is often severe enough to seriously restrict the utility of the separation process. Understanding of the operational parameters controlling the degree of polarization, and of the methods available to reduce the polarization has thus become of great economic utility, and much effort has recently been devoted to modeling of the ultrafiltration process[1,2,3,4]. Mathematical analysis of concentration polarization falls logically into three areas:

1. Diffusive ultrafiltration (or "reverse osmosis"), usually involving microsolutes in solution, and usually conducted at high pressures.

2. Ultrafiltration of macromolecules in solution (concentration, purification and fractionation of proteins, dissolved polymers, etc.)

3. Ultrafiltration of colloids.

In this section, mathematical modeling of each of these processes will be discussed separately. In each case, the objective of the modeling process is to express the dependence of the degree of concentration polarization and/or the resulting ultrafiltration flux on the properties of the solution (concentration, molecular weight, degree of purity, etc.) and on the operational variables (channel dimensions, flow velocity, pressure, temperature, properties of the membrane, etc.). In all cases, the mathematical solutions obtained refer to steady-state operation.

Concentration Polarization in Reverse-Osmosis Systems

In reverse osmosis desalination, semipermeable membranes (usually of cellulose acetate) are used which allow the permeation of liquid water without allowing salt to pass to any significant degree. The driving force for water transport is provided by hydraulic pressure more than sufficient to overcome the osmotic pressure of the salt in solution. The rejection of salt at the membrane surface then results in an increase in the salt concentration at that surface. This, in turn, causes a number of deleterious effects including a decrease in the driving force for water transport, an increase in salt leakage through the membrane, and scaling of salt crystals on the membrane surface.

Modeling of the concentration polarization phenomenon occurring in this process has been extensively investigated by many workers[5,6,7,8]. For microsolute solutions such as inorganic salts, this modeling process is quite straightforward.

The Turbulent-Flow Regime. Let us consider the case of turbulent flow of a pressurized salt solution over a salt-rejecting membrane.

A few simplifying assumptions are made first in the analysis. The membrane surface in contact with the salt solution is assumed to be smooth, convection due to local changes in liquid density is neglected, and the salt diffusion coefficient is assumed to be independent of salt concentration. Using these assumptions, the concentration polarization at the membrane surface can be described in first approximation by a simple film theory.

The film model pictures a thin, stagnant boundary layer of thickness, δ_i, between the membrane surface and the bulk solution (Figure 1). The bulk solution is considered to be well mixed, and the velocity and concentration gradients are restricted to the laminar boundary layer portion of the turbulent channel.

In the steady state, the concentration profile is constant and the overall salt transport in the stagnant boundary layer is zero. This means that the convective salt transport to the membrane is equal to the diffusive salt flux from the interface into the bulk solution. The salt balance is given by

$$J_1 \frac{dc}{dx} - D_s \frac{d^2c}{dx^2} = 0 \qquad (8)$$

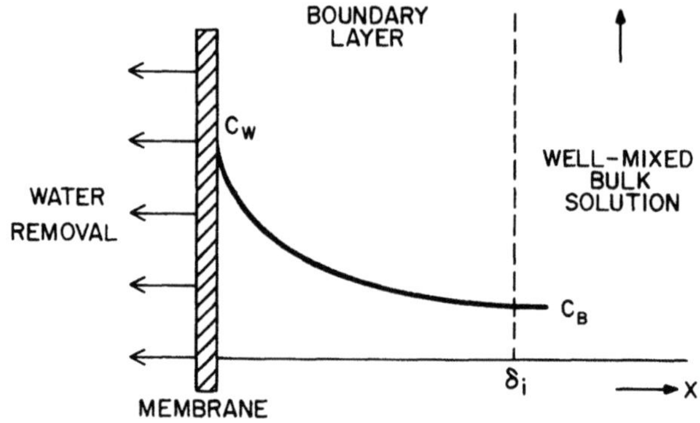

Figure 1. Concentration profile in the boundary layer for well-developed turbulent flow.

where c = salt concentration, g/cm^3; J_1 = product water flux, cm^3/cm^2 of membrane per sec; and D_s = salt diffusion coefficient, cm^2/sec. Integration of Equation (8) leads to

$$J_1 c - D_s \frac{dc}{dx} = 0 \tag{9}$$

In Equation (9), $J_1 c$ represents the solute flux toward the membrane and $D_s \left(\frac{dc}{dx}\right)$ represents solute flux away from the membrane by back diffusion. If Equation (9) is integrated once again, we obtain

$$\ln \frac{C_W}{C_B} = \frac{J_1 \delta_i}{D_s} \tag{10}$$

This equation provides a relationship between the boundary layer thickness, δ_i, and concentration of salt at the membrane surface, C_W. For salt transfer in the absence of water removal, D_s/δ_i can be replaced by k_s^o in Equation (10):

$$\frac{C_W}{C_B} = \exp \frac{J_1}{k_s^o} \tag{11}$$

It is assumed that k_s^o is unaffected by the small water flux, J_1, found in reverse-osmosis processes. Correlations of mass transfer coefficients, k_s^o for various channel geometries are available in the literature. The mass transfer in turbulent flow systems can be related to the Chilton-Colburn "j-factor"(9), which is a function of the Schmidt number.

$$j_D = \frac{k_s^o}{u_B} N_{Sc}^{2/3} \tag{12}$$

with u_B = bulk stream velocity, cm/sec; N_{Sc} = Schmidt number, $\frac{\nu}{D_s}$; and ν = kinematic viscosity, cm^2/sec. Substituting for k_s^o in Equation (11),

$$\frac{C_W}{C_B} = \exp \frac{J_1 N_{Sc}^{2/3}}{j_D u_B} \tag{13}$$

For steady turbulent flow in tubes or over flat plates, j_D is approximately equal to f/2, where f is the Fanning friction factor. With Equation (13), this leads to

$$\frac{C_W}{C_B} = \exp \frac{2 J_1 N_{Sc}^{2/3}}{f\, u_B} \tag{14}$$

Making use of the Blasius[10] correlation between friction factor, f, and Reynolds number, N_{Re}, $f = \frac{0.0791}{N_{Re}^{0.25}}$, the concentration polarization modulus, C_W/C_B, for narrow ducts in Equation (14) becomes

$$\ln \frac{C_W}{C_B} = \frac{35\, J_1\, (\nu)^{0.42}\, (h)^{0.25}}{(D_s)^{0.67}\, (u_B)^{0.75}} \tag{15}$$

where h = half channel height, cm.

The correlation of mass transfer data of a solid-liquid system in a stirred vessel is shown by Colton[19] to be

$$k_s^o = \frac{D_s}{r}\, 0.0443 \left(\frac{\nu}{D_s}\right)^{0.33} \cdot \left(\frac{\omega r^2}{\nu}\right)^{0.75} \tag{16}$$

with r = cell radius, cm; ω = stirrer speed radians/sec; and ν = kinematic viscosity, cm^2/sec. The concentration polarization modulus in a "well-stirred" batch cell is then given by substitution of k_s^o in Equation (11):

$$\frac{C_W}{C_B} = \exp \frac{J_1 \cdot r}{\left(\frac{\nu}{D_s}\right)^{0.33} \left(\frac{\omega r^2}{\nu}\right)^{0.75} \cdot 0.0443 \cdot D_s} \tag{17}$$

Although this film model is an oversimplified picture of the boundary layer transport phenomenon, it describes the concentration polarization in a turbulent flow cell and in a well-stirred batch cell with sufficient accuracy. More sophisticated models that assume the occurrence of molecular or eddy diffusion in the boundary layer lead to approximately the same result, as has been shown by Fisher[11].

Laminar-Flow Regime. Sherwood, Brian, et al.[6] have modeled the concentration polarization effect occurring in the reverse-osmosis desalination of sea water flowing in laminar flow through two dimensional channels. The underlying mechanism is, of course, identical to that shown for turbulent flow above – rejection of salt at the membrane surface results in a higher concentration of salt at the membrane than is present in the bulk of the solution, and reduces the driving force for water transport through the membrane. The modeling process is similar to that shown above for the turbulent flow region. The concentration-polarization modulus has been numerically solved by Sherwood, Brian, Fisher, and Dresner as a function of the flow geometry and permeation flux for constant flux through the membrane. Brian(5) also numerically solved for the polarization modulus allowing for a decline in flux with increasing polarization and for incomplete salt rejection in two-dimensional channels. Closed-form solutions were also obtained for some regions of operation. Near the entrance region of the channel formed by two parallel plates the approximate solution is given by Dresner(12) as

$$\frac{C_W}{C_B} = 1 + \xi + 5 \ [1 - \exp(- \sqrt{\xi/3})] \tag{18}$$

$$\xi = \frac{V_W^3 \ h \ L}{3u_{(o)} D_s^2} \tag{19}$$

where $u_{(o)}$ = velocity of fluid entering channel, cm/sec; h = half the channel height (cm); V_W = the wall permeation velocity, cm/sec; and L = the channel length, cm. Far downstream in the channel the solution reduces to

$$\frac{C_W}{C_B} = 1 + \frac{V_W^2 \ h^2}{3D_s^2} \tag{20}$$

Examination of these equations leads to the conclusions that, in laminar flow, the concentration polarization can be decreased (and the efficiency of desalination increased) by

(1) Decreasing the channel height

(2) Increasing the inlet velocity (in the entrance region)

(3) Decreasing the channel length (so as to stay in the entrance region).

The correctness of the above solutions has been experimentally confirmed for both turbulent and laminar flow conditions(11).

Concentration Polarization in the Ultrafiltration of Macromolecules in Solution

The phenomenon of concentration polarization occurring when colloidal or macromolecular solutions are ultrafiltered is governed by the same mass-transport and fluid-mechanical factors as have been shown to control microsolute polarization, but the consequences of macrosolute polarization are strikingly different. The peculiar effects of macrosolute or colloid polarization on ultrafiltration are in large measure attributable to the unusual properties of concentrated solutions of these substances, and a brief discussion of these properties will be helpful in explaining their anomalous behavior.

Rheology of Concentrated Macrosolute Solutions and Colloidal Dispersions. If one prepares solutions of polymers or dispersions of colloidal particles of increasing solids concentration, one usually observes an increase in solution-viscosity with concentration; above a certain concentration (characteristic of the particular solute), the solution ceases to behave as a Newtonian fluid and displays properties characteristic of a solid. In some cases, these concentrates behave as "true gels" – that is, they distort elastically under shear and rupture at a characteristic shear stress; in others, they behave as "Bingham-pseudoplastic" bodies – that is, they deform elastically at low shear stresses but, above a critical stress (the "yield stress"), they begin to flow as highly viscous fluids.

If concentrated solutions of such solutes are produced by polarization during ultrafiltration, there is formed adjacent to the membrane upstream surface a layer of material of substantially different rheological and mechanical properties from that of the overlying (more dilute) solution. This layer of concentrate is normally recognizable as a "slime" or "cake" adhering to the membrane surface. Because of its high viscosity or solidity, it either does not migrate with the flowing up-stream fluid, or else it moves with a substantially reduced velocity relative to that of the overlying fluid. For all practical purposes, therefore, the adherent layer or cake behaves as if it were a part of the membrane, and any solute or solvent which passes through the membrane must pass through it. It is thus important to understand, and to be able to predict, the effects which this layer has upon ultrafiltration flux and retention, and how parameters such as solute concentration, ultrafiltration pressure, and fluid shear in the feed-channel influence this layer and its transport properties.

The concentration level at which a particular macrosolute solution or colloidal dispersion begins to display anomalous rheological properties depends primarily upon the size, shape, and degree of hydration or solvation of the solute particles. For flexible-chain linear macromolecules which are well-solvated, gel-like properties may begin to develop at concentrations of 2 to 5 percent by weight. For rigid-chain, solvated macromolecules like the polysaccharides, gelation may ensue at concentrations well below 1 percent by weight, if their molecular weight is high. Highly-structured spheroidal macromolecules, such as proteins and nucleic acids, frequently can be concentrated to 10-30 percent by weight before gel-like properties are observed. With colloidal dispersions, in which the particles are unsolvated by the suspending fluid, the concentration for onset of gelation depends upon particle size and shape and the tendency to agglomerate; for submicron pigment- or mineral-dispersion, gelation may occur at volume concentrations from 5 to 25 percent solids; for polymer latices, solids content may often be increased to 50-60 percent by volume before the onset of gelation.

Obviously, an adherent gel-layer of concentrated solute overlying an ultrafiltration membrane will reduce the ultrafiltration flux (at a given hydrostatic pressure-difference), due to the additional hydraulic flow-resistance offered by that layer. To a good first approximation, the gel-layer-membrane laminate can be treated as two hydraulic resistances in series. Since the ultrafiltration flux of solvent in the membrane alone is given by

$$J_1 = \frac{\bar{P}_1}{t_m} (\Delta P) \tag{21}$$

it can easily be shown that, if the gel layer has a specific hydraulic permeability, P_g, and a thickness, t_g, the measured flux through the laminate is given by

$$J_1 = \frac{\Delta P}{\left(\frac{t_m}{P_m} + \frac{t_g}{P_g}\right)} \tag{22}$$

If J_1^o is defined as the ultrafiltration flux which would be observed with the given membrane in the absence of the gel layer at the given operating pressure (essentially the "pure solvent flux" through the membrane) then Equation (22) becomes

$$J_1 = \frac{J_1^o}{1 + \left(\frac{P_m}{P_g} \frac{t_g}{t_m}\right)} = \frac{J_1^o}{1 + \frac{R_g}{R_m}} \tag{23}$$

where R_g and R_m are the "flux resistances" of gel layer and membrane, respectively. Obviously, if the gel-layer resistance is much larger than the membrane resistance, the observed flux will be far lower than that expected to occur through the membrane alone.

Hydraulic Permeability of Gels, Influence of Polarization on Ultrafiltration Flux. The hydraulic permeability of a gel or concentrated dispersion of submicroscopic particles is a complex function of the solids concentration and of such variables as the size, shape, resistance, and state of aggregation of particles or molecules comprising the solid phase. For relatively concentrated dispersions of nearly isometric (spherical) particles, the hydraulic permeability for a solvent with unit viscosity can be approximated by the Kozeny-Carman relation for porous solids; viz:

$$P \simeq \frac{d^2}{180} \frac{\epsilon^3}{(1 - \epsilon)^2} \tag{24}$$

where d is the diameter of the particles, and ϵ is the "porosity" of the gel or dispersion (i.e., the volume fraction of the mixture which is liquid). The extreme sensitivity of the permeability to particle size and porosity is thus obvious.

Using Equation (24), it is instructive to compute the hydraulic permeability of a "gel" composed of one-micron particles, and one composed of 30 Å particles (representative of a typical protein molecule). If ϵ is assumed to be 0.5, the permeability coefficients for the two cases are, respectively, 3×10^{-11} and 2×10^{-16} cm^2. One-micron thick layers of these two gels, if permeated by water at a pressure difference of 100 psi, would display fluxes of approximately 200 and 0.0013 $cm^3/(cm^2)(sec)$. Hence, it is evident that gel layers formed by polarization of relatively large-particle size dispersions will have a negligible effect on ultrafiltration rate, whereas layers formed from macromolecular solutes may very markedly reduce the flux. Evidence in support of these expectations will be presented later in this paper.

Macrosolute Polarization Modeling and Analysis. In microsolute ultrafiltration, the pure water flux J_1 is usually an independent variable which can be changed by changing the hydrostatic pressure. The new values of J_1 then result in a new steady-state value of the dependent parameter C_W/C_B. In the macrosolute ultrafiltration, the process by which solvent flux is restricted is much different from osmotic back pressure. Since the molecular weights of the solutes are extremely high, the osmotic back pressures are too small to cause any significant reduction in the solvent activity gradient across the membrane. However, as discussed above, the concentrated solutions of macromolecules at the membrane surface which result when concentration-polarization occurs, provide a physical barrier to solvent and solute transport, acting as an additional membrane in series with the original membrane. Mathematical modeling of the flux reduction caused by this polarization is divided into two regions: (1) where the concentration polarization modulus, C_W/C_B, is low enough that the wall concentration is lower than C_g, the gel concentration of the macrosolute, and (2) when C_W/C_B is such that C_W is equal to C_g – a region which we will term "gel polarized". Since most ultrafiltration operations of interest take place in this "gel polarized" region, its modeling will be considered first, immediately following.

The gel formation process begins during ultrafiltration when, initially, the rate of convective transport of the solute to the membrane is larger than the back-transport rate. This results in an increase in the concentration of the species at the surface, until a limiting concentration is reached when further concentration cannot take place because of the impermeability of the polarized macromolecular layer to other macromolecules. (Whether this polarized layer is a gel rather than a very viscous liquid is open to debate, but experimental evidence indicates that it approaches a "close-packed" configuration of low hydraulic permeability.) This limiting value of C_W is what we have referred to as C_g, the gel concentration. Once C_W reaches C_g, further build-up of the solute must occur by thickening of the gel layer rather than by further concentration. The layer of gel so formed, acting as a hydraulic barrier in series with the membrane, reduces the solvent flux until the point is reached at which the reduced convective forward transport of the solute is balanced by diffusive back-transport of solute from the concentrated gel layer into the bulk solution. This steady state is normally reached in less than a minute, as shown by the invariance of ultrafiltration rates over short time periods.

Although the same equation [Equation (9)] relates the solvent flux J_1 to the wall concentration in both microsolute and macrosolute ultrafiltration, it is to be noted that in the former (microsolute filtration), $\frac{C_W}{C_B}$ adjusts itself to any imposed J_1 while in the latter, the wall concentration, C_W, is a constant, equal to C_g, and J_1 is a dependent variable which is constant for a given C_B and k_s. J_1 is then given by the equation

$$J_1 = k_s \ln \frac{C_g}{C_B} \tag{25}$$

Before the application of this model to specific flow geometries is discussed, a number of fairly subtle implications resulting from it should be noted. In the first place, the final ultrafiltration rate at steady state is shown to be controlled by the rate at which the retained species can be transferred **from** the membrane surface **back** into the bulk fluid. Thus, operational variables which aid in back-transport of concentrated species from the membrane (such as more rapid stirring, higher shear rates at the membrane surface, or increased temperature to aid molecular diffusion) will directly increase the final ultrafiltration rate.

It should be noted, however, that variables which simply increase initial ultrafiltration rates without providing a compensating mechanism to increase the rate of back-transport will not result in an increased steady-state ultrafiltration rate. Thus, an increase in transmembrane pressure drop, which provides an increased driving force for ultrafiltration but does not aid back-transport, would simply result in the buildup of a thicker or denser cake of retained species, the steady-state ultrafiltration rate would be reduced to its initial value, and Equation (25) would be once again obeyed. This effect is illustrated in Figures 2 and 3 with data from the ultrafiltration of cellulose waste solutions and clay suspensions and in the higher pressure regions of Figures 4 and 5 showing data from the ultrafiltration of bovine serum albumin and human plasma. The invariance of ultrafiltration rates with pressure in these regions is as predicted from the above mechanism.

Similarly, the use of a more open membrane with lower resistance to ultrafiltrate flow would not result in a net increase in the final ultrafiltration rate; the layer of retained macromolecules, again, simply grows thicker, reducing the steady-state ultrafiltration value to where it could be balanced by the back-transport rate. The invariance of plasma ultrafiltration rates with membrane permeability shown for the membranes in Figure 6, thus becomes understandable.

The model discussed above allows the development of quantitative relationships describing the dependence of ultrafiltration rates on the physical parameters of the ultrafiltration system. These relationships would quantitatively predict the ultrafiltration flux, J_1, in terms of size and concentration of the retained macromolecules in solution, the shape and dimension of the flow system and the flow rate across the membrane. As Equation (25) governs the maximum ultrafiltration flux under gel-polarization limited conditions, it can be predicted by estimating the mass transfer coefficient under any given set of geometry and fluid mechanical conditions.

The Graetz or Leveque solutions[15,16] for convective heat transfer in laminar flow channels, suitably modified for mass transfer, adequately characterize mass transfer coefficients of retained species away from the surface of the membrane. Furthermore, when diffusivities are very low, the concentration polarization boundary layers are much smaller than the channel separation for axial distances small compared with the "entrance length". It can be shown that under such conditions (usually applicable to macromolecular solutions where diffusivities are 10^{-6} cm^2/sec) the rigorous solution for the prediction of the mass transfer coefficients can be approximated by the Leveque solution

$$k_s = B \left[\dot{\gamma}_w \frac{D_s^2}{L} \right]^{1/3} \qquad (26)$$

where B is a constant dependent on the wall boundary condition, $\dot{\gamma}_w$ is the fluid shear rate at the membrane surface, and L is the length of the flow channel over the membrane.

Combining Equations (25) and (26) results in the following relationship:

$$J_1 = B \left(\ln \frac{C_g}{C_B} \right) \left[\dot{\gamma}_w \frac{D_s^2}{L} \right]^{1/3} \qquad (27)$$

Thus the water flux in the gel-polarized region would be expected to vary directly as the cube root of the wall shear rate and inversely as the cube root of the channel length. Data fitting these correlations is shown in Figure 7. The flux would also be expected to vary proportionately with the logarithm of the bulk concentration, with a plot of J_1 vs. log C_B,

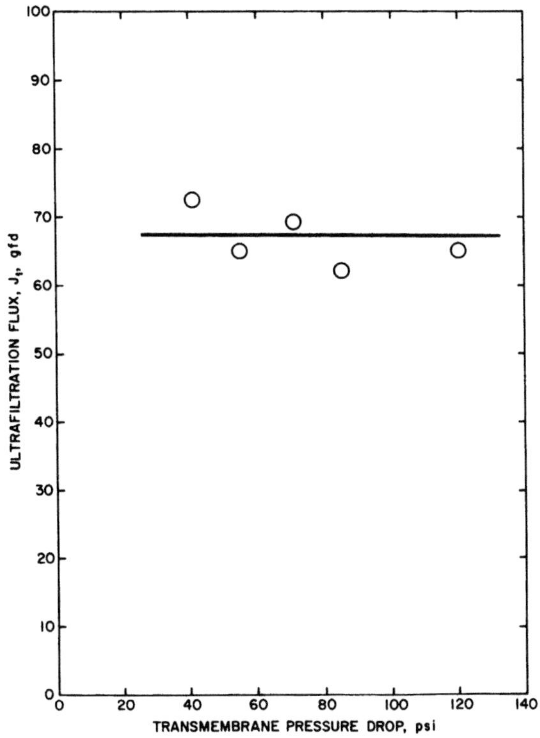

Figure 2. Dependence of ultrafiltration rate on pressure drop across the membrane. Data are for a dilute hemicellulose-caustic solution using a Diaflo® XM-50 membrane in a laminar-flow thin-channel unit.

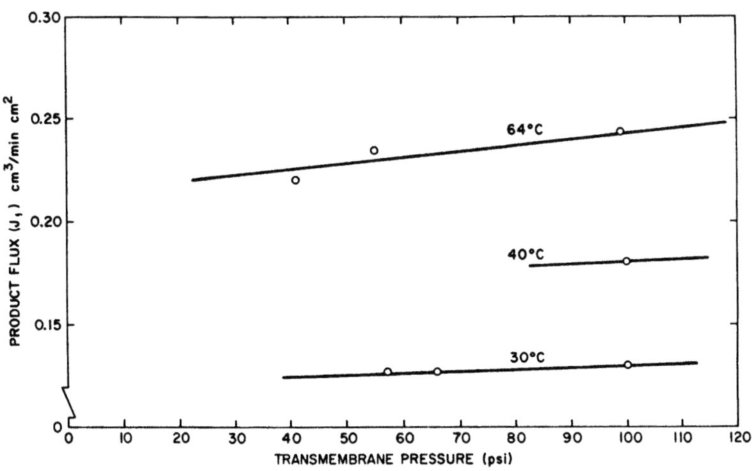

Figure 3. Dependence of ultrafiltration rate on pressure. Data were obtained with a 6 percent clay suspension in a thin-channel turbulent-flow recirculation system.

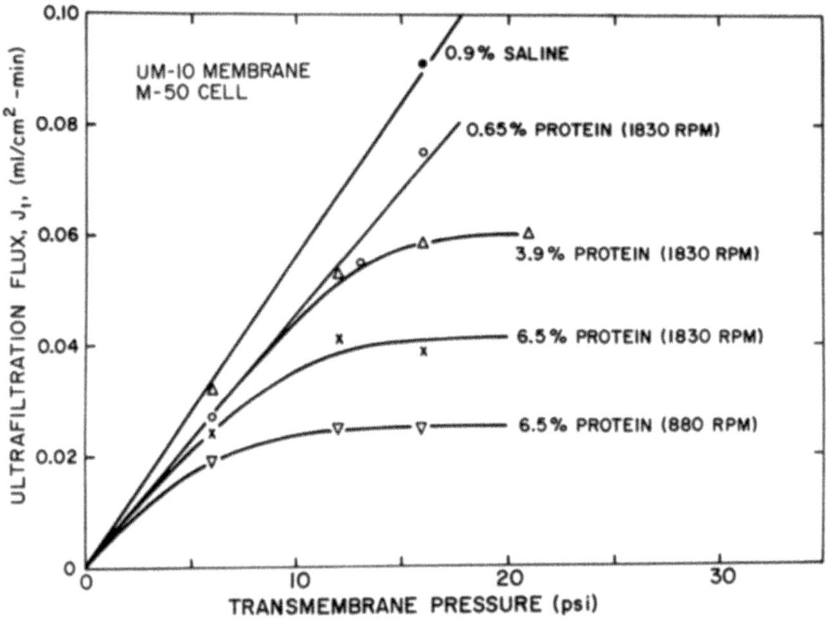

Figure 4. Flux-pressure relationships for bovine-serum albumin solutions in a stirred batch cell.

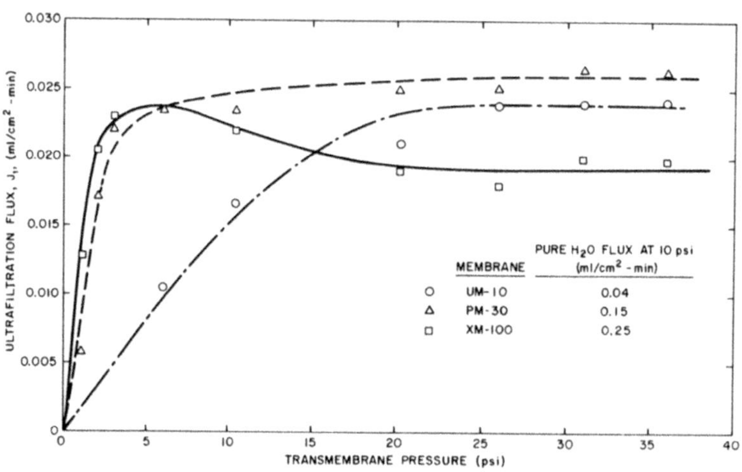

Figure 5. Flux-pressure relationships for the ultrafiltration of human plasma in a stirred batch cell using membranes of differing water permeabilities.

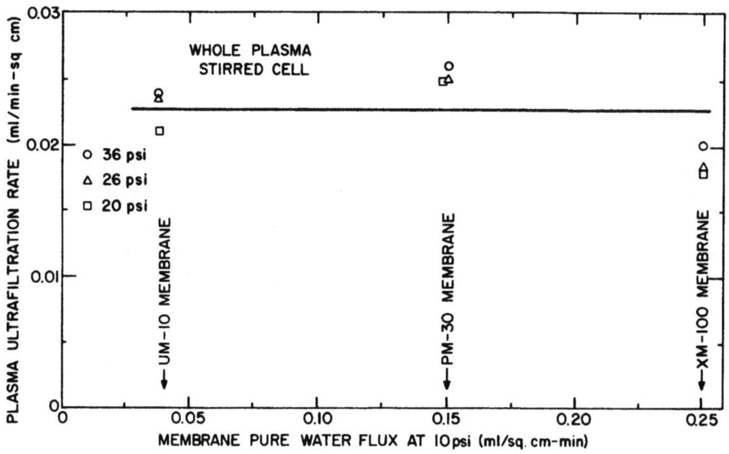

Figure 6. Dependence of ultrafiltration rates on membrane permeability at several pressures. Data are for human plasma in a stirred batch cell.

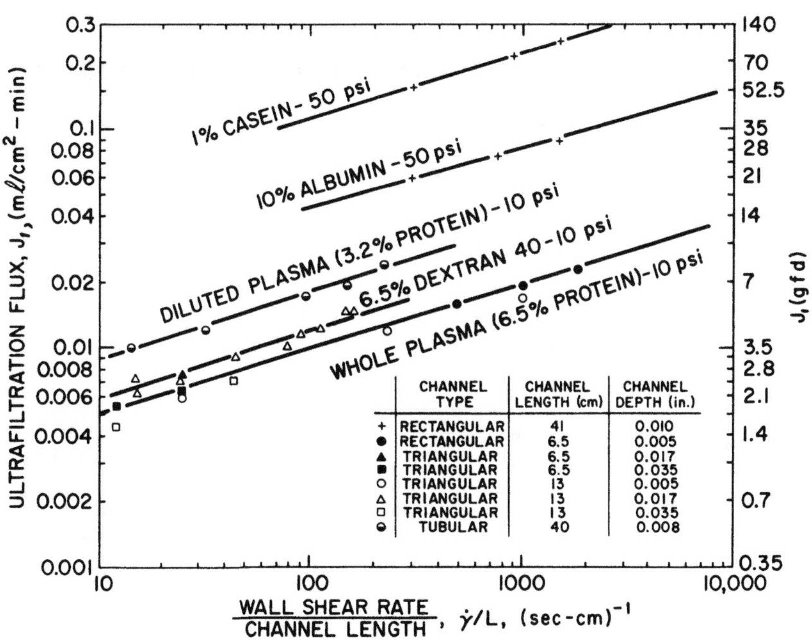

Figure 7. Ultrafiltration rates in the gel-polarized region – dependence on wall shear rate and channel length. Data are for various macromolecular solutions in thin-channel laminar-flow recirculating cells.

intercepting the log C axis at a concentration corresponding to C_g. Data plotted in this fashion are shown in Figures 8 and 9. The intercepts appear physically reasonable.

It can be seen from this relationship that the shear rate at the membrane surface is the major depolarizing parameter. Also, Equation (27) shows that operation with a number of short parallel channels is to be preferred to operation with one single long channel.

Although in very long channels Equation (26) no longer adequately describes the mass transfer solution, it can be shown that for all channel lengths of practical interest, this solution is valid. The length of the "concentration entrance region" (where this cube-root relationship can be expected to hold) is approximately 0.1 $\dot{\gamma}_w h^3/D_s$. For polymers in solution, with $D_s < 10^{-6}$ cm^2/sec and under usual operating conditions in a well-designed unit, $\dot{\gamma}_w$ will exceed 1,000 sec^{-1}, and the minimum channel depth employed in practical systems is usually about 10 mils. The entrance length is therefore at least 240 cm, certainly longer than any channel of interest in presently contemplated equipment design. This ensures the validity of Leveque solution used above for analysis of these systems.

Under gel-polarized conditions of concentration polarization, the solute concentration at the membrane surface stays constant at C_g, and the appropriate boundary condition is that of constant wall concentration, for which B takes the value of 0.816. $\dot{\gamma}_w$ depends on the flow geometry; values for three typical cross sections are listed in Table 1.

TABLE 1. RELATIONSHIP OF WALL SHEAR RATE TO FLOW GEOMETRY

Channel Shape		$\dot{\gamma}_w$
Rectangular slit	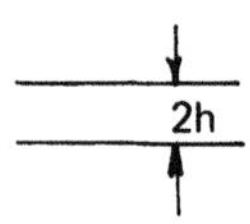	$\frac{3u_B}{h}$
Circular tube	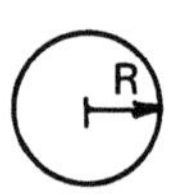	$\frac{4u_B}{R}$
Triangular channel (membrane along QR)	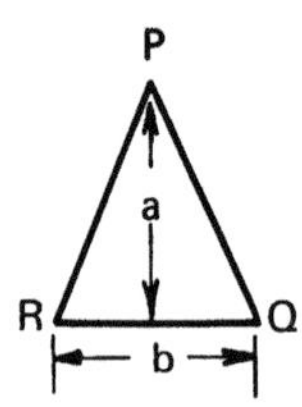	$\frac{30u_B}{a}\left(\frac{\left(5\frac{b}{a}\right)^2+12}{\left(27\frac{b}{a}\right)^2+20}\right)$

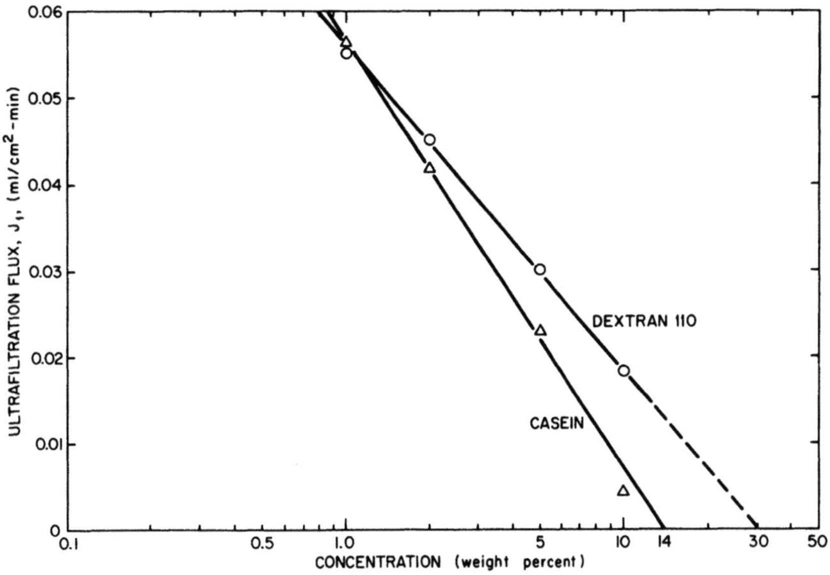

Figure 8. Decline in ultrafiltration rates with increasing concentration for solutions of casein and 110,000-molecular-weight Dextran. Data were obtained in thin-channel recirculating flow cells.

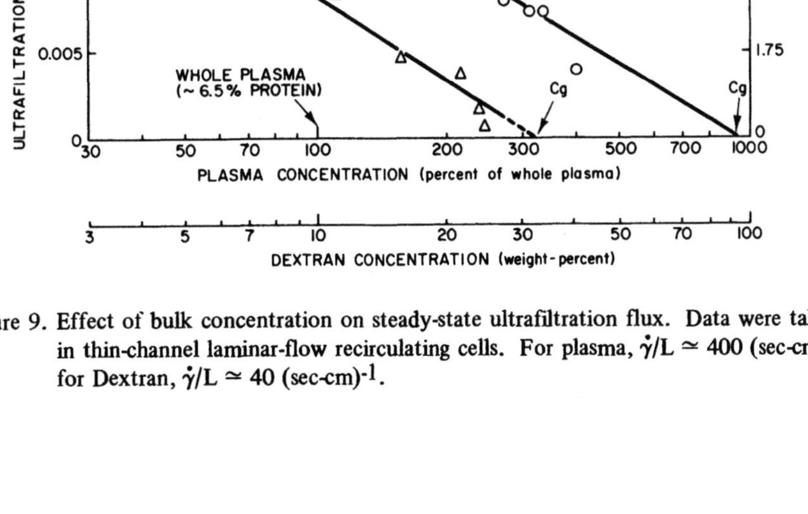

Figure 9. Effect of bulk concentration on steady-state ultrafiltration flux. Data were taken in thin-channel laminar-flow recirculating cells. For plasma, $\dot{\gamma}/L \simeq 400$ (sec-cm)$^{-1}$; for Dextran, $\dot{\gamma}/L \simeq 40$ (sec-cm)$^{-1}$.

Thus, irrespective of the channel geometry, the limiting ultrafiltration flux should vary as $u_B^{1/3}$ and $(hL)^{-1/3}$ when the above model is valid.

The theoretical model presented above requires determination of the mass-transfer coefficient, k_s, in order to estimate the maximum ultrafiltration flux under the conditions of gel polarization. The mass-transfer coefficients for turbulent flow of fluids in narrow parallel channels conventionally are written

$$\frac{k_s d_h}{D_s} = A\, N_{Re}^{a}\, N_{Sc}^{b} \tag{28}$$

where

d_h = equivalent diameter of the channel

$= 4\,\dfrac{\text{cross-sectional area}}{\text{wetted perimeter}}$

$N_{Re} = \dfrac{d_h u_B \rho}{\eta}$ = Reynolds number

u = average velocity of the fluid along the channel

ρ = density of the fluid

η = viscosity of the fluid

$N_{Sc} = \dfrac{\eta}{\rho D_s}$ = Schmidt number

A, a, and b are empirically determined constants. Three typical correlations use the values given in Table 2.

According to the model, the ultrafiltration flux should vary as u_B^a and as $D_s^{(1-b)}$. Since the concentration profile develops very rapidly in turbulent flow, one does not expect any dependence on the length along the channel.

Colton[19] has made an extensive study of mass transfer to a stationary supporting surface in an agitated cylindrical vessel. He has proposed the following correlations:

TABLE 2. TYPICAL TURBULENT-FLOW CORRELATIONS FOR FLOW IN CHANNELS

Correlation	A	a	b
Calderbank-Young[17]	0.082	0.69	0.33
Chilton Colburn[9]	0.023	0.80	0.33
Chilton Colburn[18]	0.040	0.75	0.33

(1) Laminar boundary layer over the membrane surface:

$$\frac{k_s r}{D_s} = 0.285 \left(\frac{\omega r^2}{\nu}\right)^{0.55} \left(\frac{\nu}{D_s}\right)^{0.33} \tag{29}$$

when $8{,}000 < \left(\frac{\omega r^2}{\nu}\right) < 32{,}000$

(2) Turbulent boundary layer over the membrane surface:

$$\frac{k_s r}{D_s} = 0.0443 \left(\frac{\omega r^2}{\nu}\right)^{0.75} \left(\frac{\nu}{D_s}\right)^{0.33} \tag{30}$$

when $32{,}000 < \left(\frac{\omega r^2}{\nu}\right) < 82{,}000$

where

r = cell radius, cm
ω = stirrer speed, radians/sec
ν = kinematic viscosity, cm^2/sec
D_s = solute diffusivity, cm^2/sec

Figure 10 shows the laminar region data for the ultrafiltration of 2.5 percent albumin solution in saline water. The slope of 0.42 as against 0.55 predicted by Colton is not unreasonable in view of the accuracy of data and nature of diffusing species. Figure 11 shows the turbulent region data for the ultrafiltration of bovine serum. The 0.7 power dependence is in good agreement with 0.75 power predicted by Colton.

This concludes the analysis of the gel-polarized region. The following paragraphs contain an analysis of the "pre-gel polarization" region.

The gel-controlled polarization model predicts that the ultrafiltration flux from macromolecular solutions should be independent of applied pressure – an increased pressure merely resulting in a thicker gel layer that then retards the flux down to its original value where the solute flux can then be balanced by back diffusion. Although this pressure independence indeed has been shown for many systems, particularly at high pressures, notable deviations have occurred in some systems at low pressures. A typical example is shown in Figures 12, 13, and 14, with data taken on bovine serum albumin in a thin-channel, laminar-flow cell.

It is hypothesized that this region is one in which the concentration polarization modulus $\left(\frac{C_W}{C_B}\right)$ is so low that the wall concentration, C_W, is below the gel concentration, C_g. It is also thought, however, that the wall concentration is nonetheless high enough to provide a concentration boundary layer with a significant physical barrier to water transport. Just as in the cases of microsolutes and of macrosolutes in the gel-polarized region, the final resulting ultrafiltrate flux is determined by a balance of forward-convective transport of solute and back-diffusive

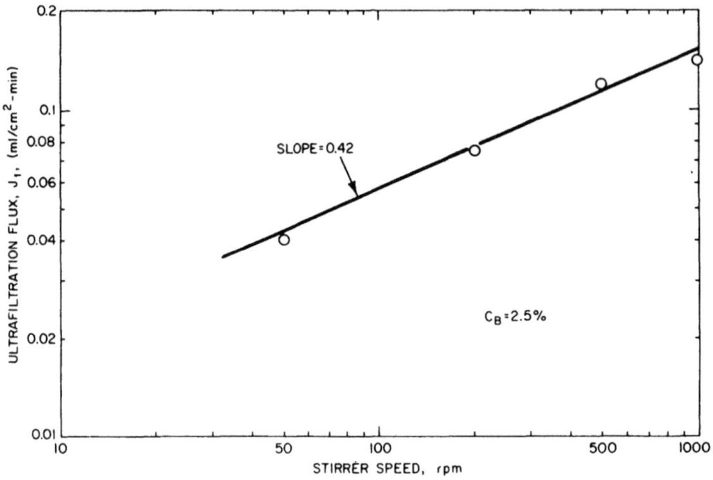

Figure 10. Dependence of ultrafiltration flux on stirrer speed in a laminar flow stirred batch cell. Data are for a 2.5 percent human-serum albumin solution in a Diaflo® Model 50 cell at 50 psi.

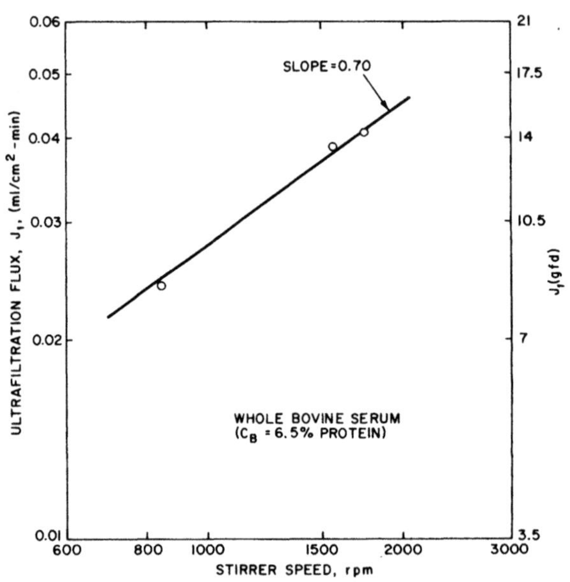

Figure 11. Dependence of ultrafiltration flux on stirrer speed in the high-speed region. Data are for 6.5 percent bovine-serum albumin.

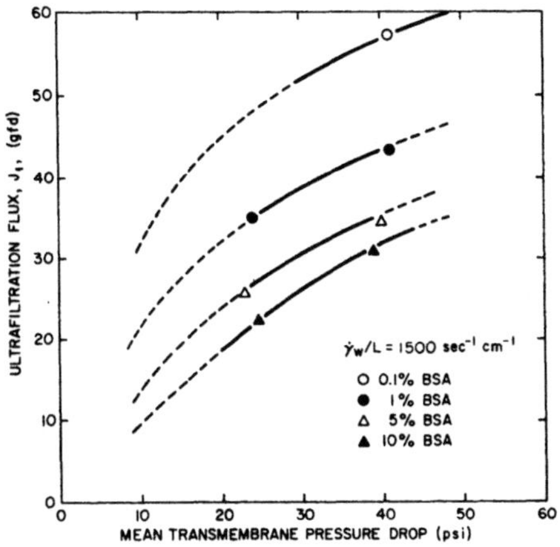

Figure 12. Ultrafiltration of bovine-serum albumin in a thin-channel recirculating system – effect of pressure. Data were obtained in a Diaflo® TC-1 cell with $\dot{\gamma}/L$ = 1500 (cm-sec)$^{-1}$.

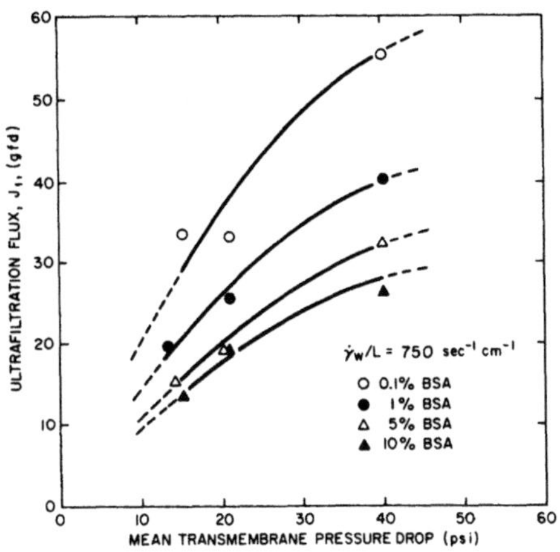

Figure 13. Effect of pressure on the ultrafiltration rates of solutions of bovine-serum albumin in a TC-1 thin-channel recirculating system; $\dot{\gamma}/L$ = 750 (cm-sec)$^{-1}$.

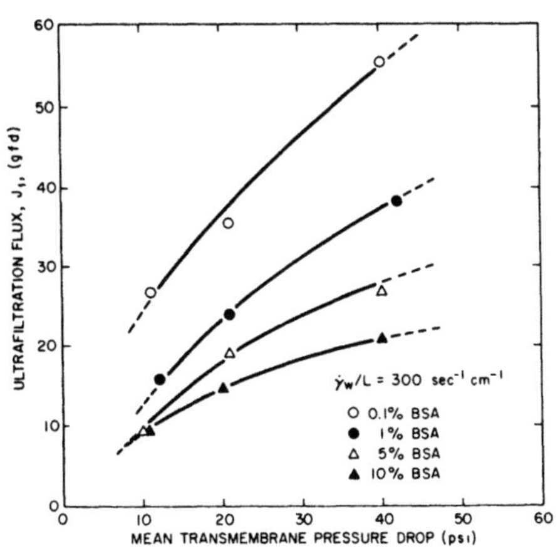

Figure 14. Effect of pressure on the ultrafiltration rates of solutions of bovine-serum albumin in a TC-1 thin-channel recirculating system; $\dot{\gamma}/L$ = 300 (cm-sec)$^{-1}$.

transport. For example, as the pressure is increased, more solute is brought to the membrane surface and C_{wall} presumably increases, thereby increasing the back-transport rate of solute into the flowing stream. Eventually, as the pressure is further increased, the concentration polarization modulus increases, and C_{wall} approaches C_g. No further increase in the concentration-gradient driving force for back diffusion can then take place, and the ultrafiltration flux henceforth remains constant. This hypothesis is supported to some degree, at least, by the concentration dependence of the pressure effect: at higher bulk concentrations (when C_{wall} would be expected to approach C_g at lower pressures), the slope of the flux-pressure curve is smaller. Thus, the flux appears to be approaching a pressure-independent value at lower pressures with high concentration solutions than with more dilute solutions.

Similar pressure-dependent data are shown in Figures 5 and 15 for whole plasma. In Figure 5, the width of the pressure-dependent region is shown to be a function of the membrane permeability: the fluxes through the higher water-flux membranes become independent of pressure at much lower pressure than those through the lower permeability membranes. This finding is consistent with the hypothesized "pre-gel polarization" model: at any given pressure, the higher initial transport rates through the looser membranes cause a higher concentration polarization modulus (higher ratio of C_{wall} to C_{bulk}), so that C_{wall} approaches C_g at lower pressures. In Figure 15, the width of the pressure-dependent region increases directly with the fluid shear rate across the membrane face. Again, this appears consistent with the hypothesis: higher shear rates result in **lower** concentration polarization moduli, and the pressure must be increased further before C_{wall} approaches C_g.

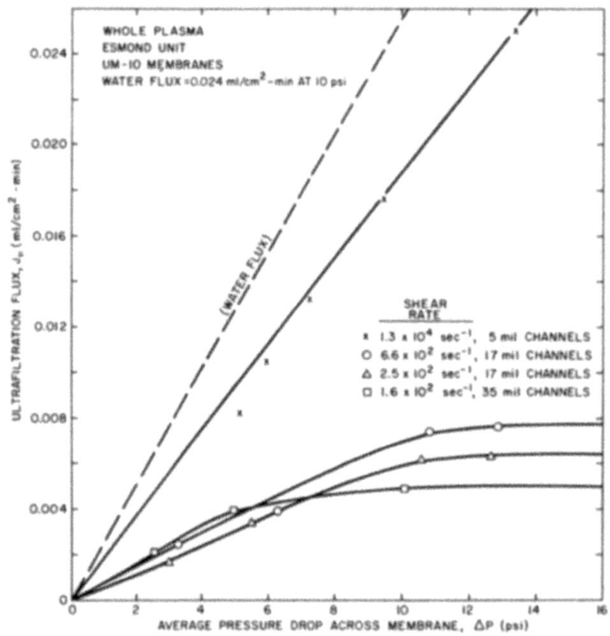

Figure 15. Dependence of ultrafiltration rates on pressure in the low-pressure region.

Data are for human blood plasma in a recirculating system with thin channels of triangular crosssection.

Mathematical modeling of this region proceeds similar to that for microsolute polarization. The polarized region is modeled as a membrane in series with the original membrane, and the resulting flux is given by

$$J_1 = \frac{\Delta P}{R_m + R_P} \tag{31}$$

where R_m is the membrane resistance and R_P is the resistance of the polarized boundary layer. R_P is, in turn, given by

$$R_P = \frac{t_b}{\bar{P}} \tag{31a}$$

where t_b is the thickness of the boundary layer and $\bar{P}$ is its overall hydraulic permeability. (Note that the permeability of the boundary layer would be expected to be a strongly decreasing function of the local solute concentration; $\bar{P}$ is therefore an average permeability, normalized over the entire boundary layer thickness, t_b. $\bar{P}$ can be expressed, at least in theory, as a unique function of $\left(\frac{1}{C_w}\right)$.

Also, as long as C_W is less than C_g, C_W is a function of the transmembrane pressure drop, increasing with increasing ΔP such that the concentration profile in the boundary layer maintains its exponential shape. (The boundary layer thickness remains constant, being a function only of the fluid mechanical conditions.) As a result, the flux increases less than proportionately with increasing ΔP. Thus, in the steady state,

$$J_1 = \frac{\Delta P}{R_m + \frac{t_b}{\bar{P}}} \tag{32}$$

Also, a solute balance at steady state gives us

$$J_1 = k_s \ln \frac{C_W}{C_B} \tag{33}$$

from (32) and (33),

$$\ln \frac{C_W}{C_b} = \frac{1}{k_s} \left[\frac{\Delta P}{R_m + \frac{D_s}{k_s \bar{P}}} \right] \tag{34}$$

or,

$$\ln \frac{C_W}{C_B} = \frac{\Delta P}{k_s R_m + \frac{D_s}{\bar{P}}} \tag{35}$$

Equation (35) reveals the following conclusions:

(1) The onset of the pressure-independent region occurs at lower pressures for higher bulk concentrations and vice versa.

(2) The onset of the pressure-independent region occurs at higher pressures for higher mass-transfer coefficients (i.e., higher fluid shear rates).

(3) The onset of the pressure-independent region occurs at higher pressures for higher membrane resistance, i.e., for tighter membranes.

Figures 11 through 15 support these conclusions qualitatively.

It has been pointed out that Equation (32) describes the flux-pressure drop relationship in the low-pressure region. For a pure solvent, $\bar{P} = \infty$, and the flux is then directly proportional to ΔP. When a solute is present, $\bar{P} = f\,(C_W)$, $= f'\,(C_B, \Delta P)$. P **decreases** with increasing ΔP and hence J_1 increases less than proportionately with ΔP. In the pressure-independent region ($C_W = C_g$),

$$J_1 = \frac{\Delta P}{R_m + \frac{t_b}{\bar{P}_o} + \frac{t_g}{P_g}} \tag{36}$$

where $\bar{P}_o$ is the boundary-layer permeability when $C_W = C_g$, t_g is the thickness of the gel layer and P_g is the permeability of the gel. At the onset of gelation, $t_g = 0$ and $\bar{P} = \bar{P}_o$. Thus, Equations (35) and (36) converge at the onset of gelation. Equation (36) governs J_1 after gelation when J_1 becomes independent of ΔP, since t_g adjusts itself to nullify the change in ΔP. In this region, $\bar{P}_o$ remains constant.

The Effects of Concentration Polarization on Retentivities. As discussed previously, the rejection, σ of a solute through a membrane is defined as

$$\sigma = 1 - \frac{C_f}{C_B} \tag{37}$$

where C_f is the concentration of solute in the filtrate and C_B is the concentration of solute in the bulk solution upstream of the membrane. For very dilute solutions, operating at such low pressures and high shear rates that the concentration polarization modulus is negligible, σ for a particular solute is a function of the membrane properties alone (e.g., average pore size and pore-size distribution, absorptivity of the membrane for the solute, etc.). However, as the operating conditions are changed and concentration polarization becomes significant, the effect of the gel layer formed by this polarization becomes equivalent to the interposition between the primary membrane and solution of a secondary membrane through which all ultrafiltrate must pass. Clearly, if the solute-retention characteristics of this secondary membrane are different from those of the primary membrane, this fact will be reflected in the composition of the ultrafiltrate.

If the feed solution contains a single solute and if the primary membrane is essentially completely impermeable to that solute, the formation of a gel layer of solute (while it may markedly reduce flux) will not influence solute retention; the ultrafiltrate will be solute free with or without polarization. On the other hand, if the primary membrane is partially permeable to the solute, the effects of polarization on solute retention will be a function of the **mechanical properties** of the gel layer. If the layer is viscous but still fluid, so that solute molecules within the layer are relatively mobile, then the locally high solute concentration near the membrane surface will cause an increase in the solute flux through the primary membrane, with attendant decrease in rejection. For example, let us assume that the solute is of uniform molecular weight and is nonaggregating (i.e., solute-solute interaction is negligible) and that the gel layer remains mobile. (This situation is likely to obtain when ultrafiltration is conducted at very low pressures, with quite dilute solutions, and/or with solutes that do not form gels except at quite high solids concentration.) Let us further assume that the finite rejection coefficient is caused by a distribution of pore sizes in the membrane, some of which are large enough to pass solute and others of which are too small for solute passage.

If a fraction, a, of the total **solvent** flux passes through the solute-rejecting pores (while a fraction, $1\text{-}a$, carries solute at a concentration C_B through the membrane), the rejection coefficient at infinite dilution (or at a concentration polarization modulus, C_W/C_B, of 1.0) is given by

$$\sigma = 1 - \frac{(1\text{-}a)\, C_B}{C_B} = a \quad . \tag{38}$$

However, as the concentration at the membrane surface is increased above the bulk concentration by the concentration polarization mechanism, the solute that is carried through the membrane is then at a higher concentration C_{pore} (where $C_W \geqslant C_{pore} > C_{bulk}$ and C_{pore} increases with increasing C_W). σ is then given by

$$\sigma = 1 - \frac{(1\text{-}a)\, C_{pore}}{C_B} \quad . \tag{39}$$

Thus, σ decreases as the concentration polarization modulus (and therefore, C_W) increases. This effect is illustrated in Figure 16, which presents data from ultrafiltration experiments on dextran solutions of various molecular weights in a stirred cell.[21] The rejection coefficient can be seen to decrease with increasing pressure (as would be expected in the pre-gel polarization region, since increasing pressure increases the concentration polarization modulus).

Similar effects were encountered when defatted milk whey was fractionated using a series of Diaflo® ultrafiltration membranes. Defatted milk whey is a complex mixture of proteins of diverse molecular weights, as shown in Table 3. Upon fractionation of this mixture through a cascade of membranes with differing molecular-weight cutoffs, it was found that the lowest molecular-weight fractions (<10,000) of the mixture passed through membranes with higher cutoffs, unimpeded by the polarization layer on the membrane surfaces. The medium molecular-weight protein fractions, however, were greatly hindered by the polarized layer and were unable to pass through even the most open membrane. Data from the run are shown in Tables 4a and 4b. Comparison of actual and theoretical yields in Table 4b show that protein fractions C, D, and E, which would have been expected to pass through the XM-50 membranes, were instead largely retained by the membrane, presumably because of the polarized layer formed above this membrane by the higher molecular species of fractions A and B.

TABLE 3. ANALYSIS OF WHEY (DEFATTED) STARTING SOLUTION

Estimation of the molecular weights from the elution patterns following chromatography of Sephadex G-100. (Method of Whitaker)

Component	V	V/V_0	M.W.	Weight % in Mixture	Identification
A	15.5 (= V_0)	1.00	>100,000	12.5	? Macroglobulin
B	20.5	1.37	70,500	5.1	Albumin
C	27.5	1.83	28,800	26.3	β-lactalbumin
D	32.8	2.18	14,600	11.8	α-lactalbumin
E	45.5	2.83	~5,000	4.1	Peptones
F	50.8	3.27	~2,000	36.8	Peptones
G	57.0	3.70	<1,000	3.3	Amino acids, etc.

TABLE 4a. WHEY ANALYSIS

Fractional distribution analysis of the membrane-separated components.

Membrane Compartment	Nominal Membrane Compartment Molecular Weight Range[a]	Total Protein, mg	Fractions Retained in Compartment, %						
			A	B	C	D	E	F	G
> XM-50	>50,000	786	32.5	4.8	40.7	22.1	–	–	–
PM-10 – XM-50	10,000-50,000	39	22.8	2.6	11.6	37.8	3.3	15.4	6.4
UM-2 – PM-10	2,000-10,000	180	4.7	–	–	7.1	11.3	70.1	6.8
< UM-2	<2,000	606	1.6	–	–	–	–	96.7	1.8

TABLE 4b. COMPARISON OF ACTUAL AND THEORETICAL YIELDS (FROM MOL WT) OF TOTAL PROTEIN ON A PERCENTAGE BASIS

Nominal Membrane Compartment Molecular Weight Range[2]	Actual Protein Yield, %	Theoretical Yield, %
>50,000	49.7	17.6
10,000-50,000	2.4	38.1
2,000-10,000	11.2	4.1
>2,000	37.6	40.1

(a) Based on known Diaflo membrane retention characteristics for single solutes in solution.

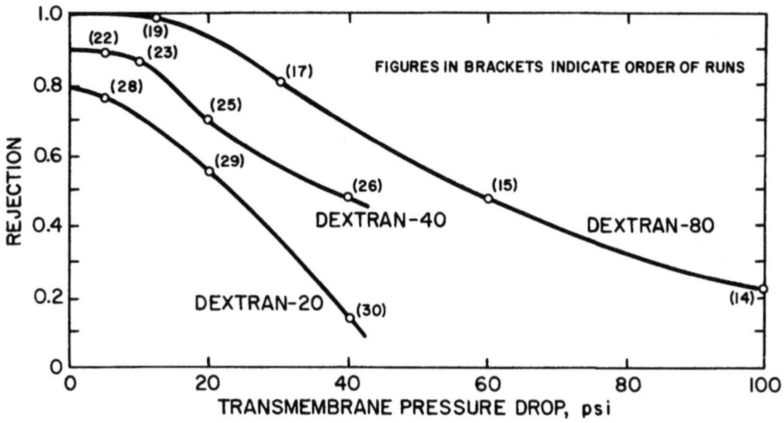

Figure 16. Effect of pressure on the rejection of Dextran solutions of various molecular weights. Data were obtained with a stirred batch cell using a Diaflo® XM-50 membrane and a bulk concentration of 1 percent Dextran.

Converse to the phenomenon discussed above, if the polarized layer is coherent and truly gel-like so that molecules within the layer are essentially immobile (a situation encountered at high pressure, with concentrated solutions, and with solutes which are strong gel formers) it becomes virtually impossible for solute molecules to pass through the layer and reach the primary membrane. In such a case, the gel layer serves as a partial or complete barrier to passage of solute, and the retention efficiency increases.

The influence of polarization and gel-layer formation on retention becomes most marked when the solutions being ultrafiltered contain **more than one solute** and when the primary membrane is retentive for one or more solutes yet permeable to others. In this case, the gel layer formed of the retained solute or solutes, if it is of very fine pore texture or if it is composed of molecules that interact strongly with the other solutes in the solution, may be partially or completely retentive for solutes to which the primary membrane is permeable. Under those circumstances, the smaller molecules are retained by the gel layer, and the apparent rejection by the membrane for the smaller molecules increases.

In Figure 17, data are presented on the rejection of human serum albumin ultrafiltered through XM-100 membranes in the presence of gamma globulin. Although the albumin rejection coefficient was approximately zero in the absence of γ-globulin, it rapidly increased as γ-globulin (the rejection of which is >0.95) was added to the solution. The data in Figure 17 fit the equation

$$\sigma_{Alb} = 2.4 \sqrt{\text{Conc. } \gamma\text{-globulin}} \qquad (40)$$

although there is, at present, no theoretical reason known for this square-root dependence.

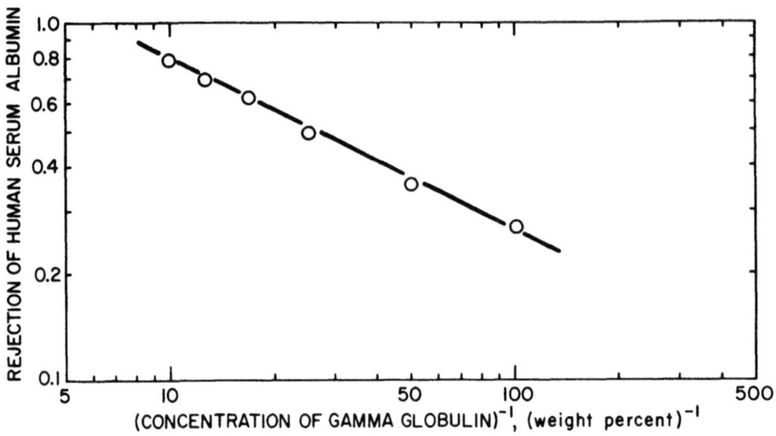

Figure 17. The interference of gamma globulin with the passage of human-serum albumin through a Diaflo® XM-100 membrane.

The impact of polarization on solute separation by ultrafiltration is even further aggravated if the membrane-impermeable solutes present possess polyelectrolyte character. These properties particularly are encountered frequently with biological polymers such as proteins, nucleic acids, and many polysaccharides. In these cases, the concentrated gel layer found by polarization may contain a relatively high-ionic-charge density, and by virtue of the Donnan ion-distribution equilibrium, excludes even simple electrolytes. Accordingly, it is often observed on ultrafiltration of such polymeric solutes that the ultrafiltrate is significantly depleted in simple electrolytes (salts, acids, bases), despite the fact that the primary membrane (in the absence of polarization) is freely permeable to such substances. While this phenomenon is undesirable in most purification applications of ultrafiltration (such as demineralization of protein solutions), it has effectively been utilized by Kraus and coworkers[13,14] for the preparation of "dynamic" reverse-osmosis membranes useful for water desalination.

It is evident that polarization eliminates or greatly reduces the separative capability of a membrane so that if the purpose of the ultrafiltration is to effect a separation between larger and smaller molecules, the efficiency of the process is seriously compromised by such polarization.

Ultrafiltration of Colloidal Suspensions

Colloidal dispersions (represented by skimmed milk, polymer latices, and clay suspensions) have been subjected to ultrafiltration in both laminar and turbulent-flow systems; typical data are shown in Figures 18 and 19. If one assumes that macrosolute polarization theory is applicable to these systems, it is possible to estimate C_g/C_B from Equation (25) by observing J_1 and estimating k_s from the fluid-mechanical conditions. When such calculations are carried out for the above materials, it is found that C_g/C_B is unrealistically high (see Figures 18 and 19, and Tables 5 and 6). In other words, the actual ultrafiltration fluxes are far higher than would be predicted by the mass-transfer coefficients estimated by conventional equations, with

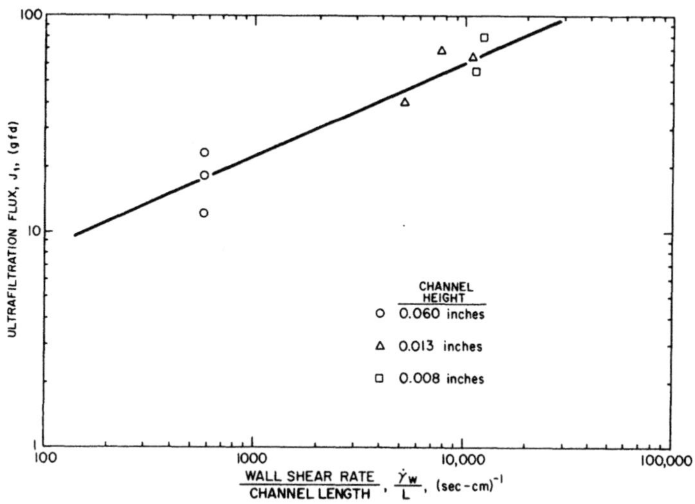

Figure 18. Ultrafiltration of a 10 percent (by weight) suspension of latex particles. Data were taken in a thin-channel recirculating system.

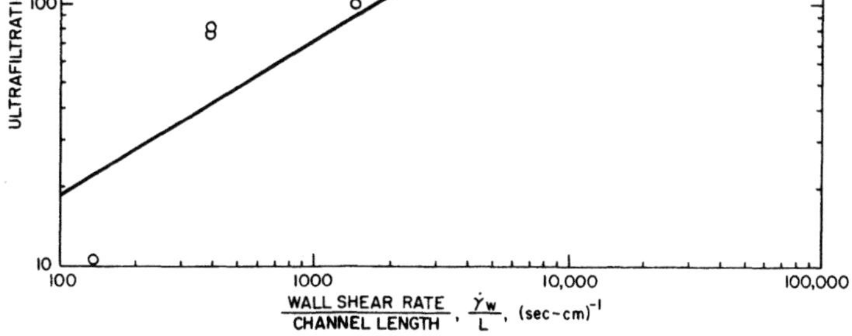

Figure 19. Ultrafiltration of a 6 percent clay suspension in a thin-channel recirculating system.

TABLE 5. THE SOLUTIONS ULTRAFILTERED

	Casein	Clay	Latex
Particle Radius, A	55	2500 to 5000	2500
Approximate Diffusion Coefficient, cm^2/sec	3×10^{-7}	1×10^{-8}	1×10^{-8}
Concentration	1%	6%	10%
Description	Threadlike protein	Mixed kaolin and montmorillonite platelets	Synthetic polymer emulsion

TABLE 6. POLARIZATION MODULI FOR ULTRAFILTRATION OF COLLOIDAL SUSPENSIONS

Suspension	k_s, cm/sec (calculated from flow conditions)	Flux, cc/cm^2-min	C_W/C_B (calculated from flux)
Latex	8.5×10^{-5}	0.19	6.3×10^{15}
6% Clay	4.7×10^{-5}	0.28	10^{63}
1% Skimmed Milk	3.47×10^{-4}	0.28	6×10^{5}
6% Clay	2.4×10^{-5}	0.028	2.2×10^{8}

the assumption that the proper diffusion coefficients are the Stokes-Einstein diffusivities for the primary particles.

These observations lead to the conclusion that either (a) the "back-diffusion flux" of particles from the polarized layer is, in fashion, substantially augmented over that expected to occur by Brownian motion or (b) the transmembrane flux is not limited by the hydraulic resistance of the polarized layer over any reasonable range of layer thickness.

Of these two possibilities, the latter appears to us to be by far the more plausible. As pointed out earlier, closely packed cakes of micron- or submicron-sized particles have quite high hydraulic permeabilities: A cake comprised of 1-micron particles 1 centimeter thick will show a permeation flux to water (at 100 psi) of 2 x 10^{-2} cm^3/cm^2 sec. As a consequence, it would appear that the removal of particles from the "polarized" particulate cake forming on the membrane with colloidal suspensions takes place by a completely different mechanism than is observed with micro or macrosolute polarization. With these suspensoid systems, particles convectively transported to the membrane are accumulated as a rapidly thickening cake layer; when the layer thickness reaches a certain (quite large) value dependent on the fluid velocity and channel height, a portion (if not all) of the cake layer is caused to flow laterally over the surface of the membrane by fluid shear and is discharged with fluid at the exit of the channel. In this situation, particle diffusion normal to the membrane surface plays no role in the polarization process, and only fluid dynamics and cake rheology govern the steady-state cake thickness. Additional experimental studies of colloid ultrafiltration will be required, however, to adequately test this hypothesis.

FLUX DECAY WITH TIME

The back-diffusion-controlled-concentration polarization model for macrosolute ultrafiltration predicts a very rapid drop in ultrafiltration flux (occurring in a few seconds or less as the gel layer is built up) followed by a flux that remains constant with time. Figure 20 shows this constant flux behavior when a 6 percent solution of 40,000-molecular-weight Dextran was ultrafiltered. (The flux decay below the water flux of the membrane occurred in an immeasurably short time and is not shown here.)

In some practical systems, however, a gradual decay in flux is noted after long-term operation. This flux decay has serious practical implications and, at least at this time, has not been sufficiently well modeled to allow actual prediction of its magnitude or methods of control. Figure 21 shows such results encountered with 1 percent casein solution over a period of 350 minutes. After the initial abrupt drop, the flux continues to fall with time at a much slower rate. This drop continues for very long times.

One probable cause of this effect is an irreversible consolidation (or "hardening") of the solute gel with time, which would then reduce the hydraulic permeability of the gel continuously with time. As noted earlier, the hydraulic permeability, P_g, can be approximated by the Carman-Kozeny equation,

$$P_g = K \frac{\epsilon^3}{(1-\epsilon)^2} , \tag{41}$$

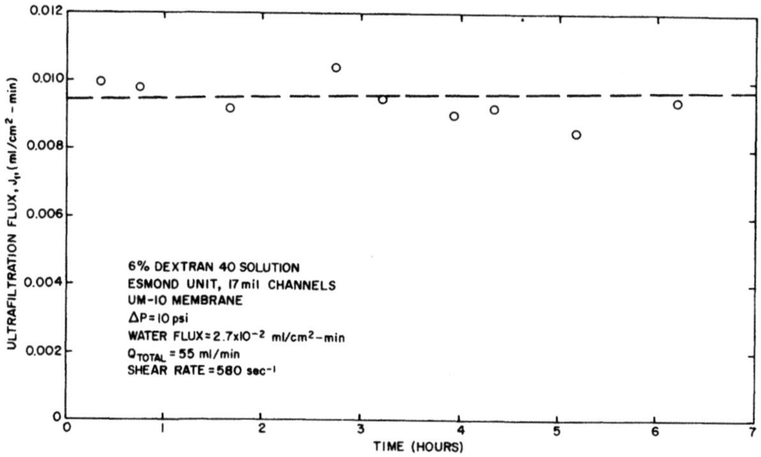

Figure 20. Variation of ultrafiltration flux with time for a 6 percent solution of 40,000 molecular weight Dextran. Data were taken in a laminar-flow recirculating system with thin, triangular channels; $\dot{\gamma}/L$ = 45 (sec-cm)$^{-1}$.

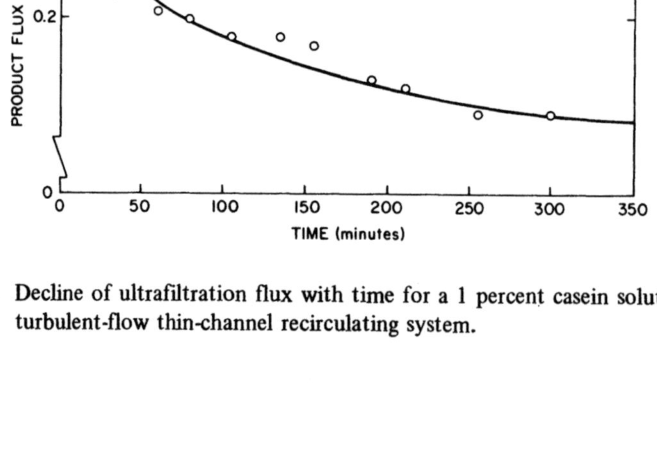

Figure 21. Decline of ultrafiltration flux with time for a 1 percent casein solution in a turbulent-flow thin-channel recirculating system.

where ϵ is the void fraction of the cake and K is a proportionality constant. At high ϵ ($\epsilon > 0.8$), P_g is proportional to $(1-\epsilon)^{-2}$, while at low ϵ ($\epsilon < 0.2$), P is proportional to ϵ^3. Thus, as ϵ falls due to prolonged exposure to high pressure, P_g drops very rapidly. Reduction in cake permeability is, of course, accompanied by a decline in transmembrane flux; as soon as the flux falls below the value limited by solute back diffusion, an increase in driving pressure will increase the flux, but never to a value in excess of that observed prior to gel consolidation.

Since the long-term time effect is a result of the compression of the gel, it should be possible to restore the initial flux by mechanically removing the gel layer. This has been confirmed for the majority of macrosolute systems so far examined, as illustrated in Figure 22, for the ultrafiltration of secondary effluent from an activated sludge plant.

ILLUSTRATIVE EXAMPLES OF USES OF ULTRAFILTRATION

Concentration of Crude-Animal Protein by Batch Ultrafiltration(23)

Ultrafiltration has been used to recover and concentrate crude proteins obtained from meat wastes. The following is an example of one such process performed in our laboratories. The protein mixture was concentrated from 3 to 20 percent total solids, after which it was diafiltered to remove low molecular salts and other solutes. For the present illustration we will only examine the ultrafiltration part of the process.

Ultrafiltration was performed in an Amicon TC-1 ultrafiltration system. With laminar flow at a shear rate of approximately 13,000 sec^{-1}, the mean ultrafiltration pressure was 50 psi. At this pressure, a UM-10 membrane would have a pure-water flux of approximately 75 gfd.

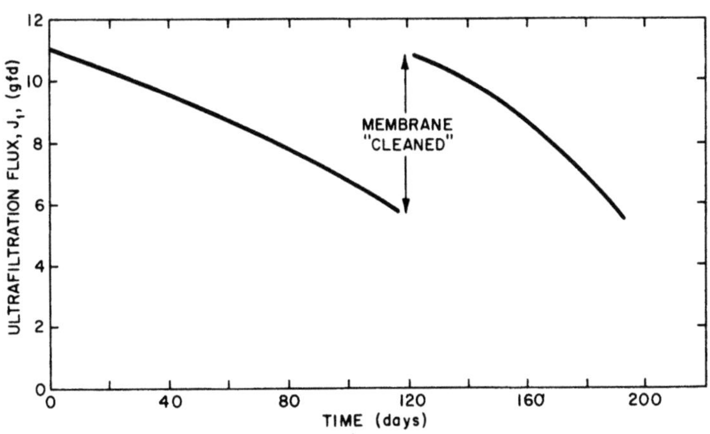

Figure 22. Ultrafiltration of a secondary effluent from an activated sludge plant at constant pressure and flow conditions over the membrane.

(Courtesy of Dorr-Oliver, Inc.)(22)

Figure 23 shows the ultrafiltration flux at room temperature as a function of solids content for this protein meal. The logarithmic decay in ultrafiltration flux with concentration is observed, but the extrapolated value of C_g is greater than 100 percent for the data obtained at room temperature. This suggests either that all of the reported solids content of the crude protein meal was not participating in the polarization process (since the crude protein contained about 30 percent salts, peptides, and low-molecular-weight carbohydrates which passed through the membrane, this explanation seems reasonable) or that operation was in the pre-gel polarized region. When the ultrafiltration is performed at 4 C, much lower fluxes and a much lower apparent value of C_g is observed. Apparently, this reflects the reduction in the mass-transfer coefficient and the greater tendency of proteins to gel at lower temperatures. The recirculation rate in these low-temperature experiments was also lower so that the effective shear rate was only 6,500 sec^{-1}. This would further reduce the mass-transfer coefficient.

Figure 24 is a plot of solids concentration versus time for the ultrafiltration of the crude protein at two sets of operating conditions. The experimental data are compared with the solids-concentration/time curves calculated making use of Equation (25) and a mass balance around a batch cell. This results in the following expression for the time, θ, required to carry out a particular ultrafiltration in terms of the initial concentration of the retained species, C_o, their final concentration, C_F, the initial volume of solution being ultrafiltered, V_o, the membrane ultrafilter area, A_m, the mass-transfer coefficient corresponding to the flow conditions over the membrane, k_s, and C_g as defined in Equation (25):

$$\theta \simeq \left(\frac{C_o}{C_g}\right)^{0.24} \left(\frac{V_o}{k_s A_m}\right) \left[1 - \left(\frac{C_o}{C_F}\right)^{0.76}\right] \qquad (42)$$

The values of k_s used in these calculations are reported in the figure and were obtained from the experimental data at the starting concentration of 3 percent solids. Use of these initial values of k_s results in a somewhat conservative calculation, as noted in Figure 24. However, considering the assumptions involved, the agreement between theory and experiment is quite good; certainly, it is satisfactory for preliminary design purposes.

Ultrafiltration of Blood and Plasma in a Model Artificial Kidney

Ultrafiltration (preceded or followed by dilution) has been proposed as an alternative process to dialysis for blood cleansing in an artificial kidney[24,25,26,27]. The success of such a process is dependent upon a membrane sufficiently discriminating to pass blood toxins (usually of molecular weights less than 1000) while retaining blood proteins and cells. The Diaflo UM-10 membrane has been shown to possess this discrimination, as shown in Figure 25. At the same time, however, successful operation depends upon achieving sufficiently high fluxes through the membrane to allow the cleansing operation to take place within limited time (usually, 6 to 10 hours) in a unit of practical size. The flux reduction caused by concentration polarization of blood proteins and the control of this polarization thus become of vital importance to the process.

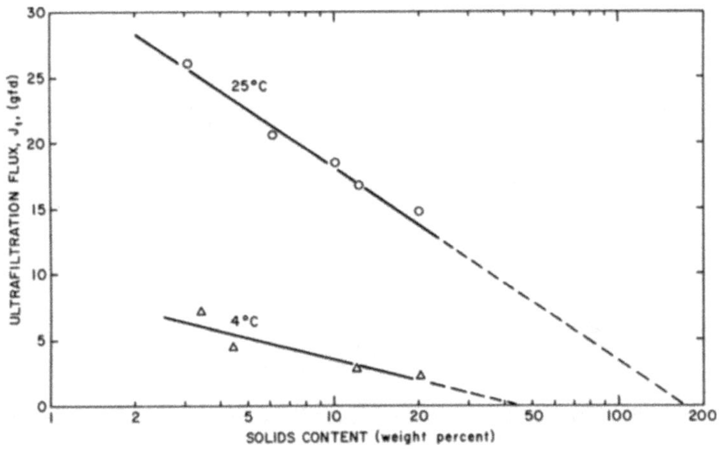

Figure 23. Ultrafiltration of a high-protein meal – variation of flux with concentration and temperature. Data taken in a thin-channel, laminar-flow recirculating system.

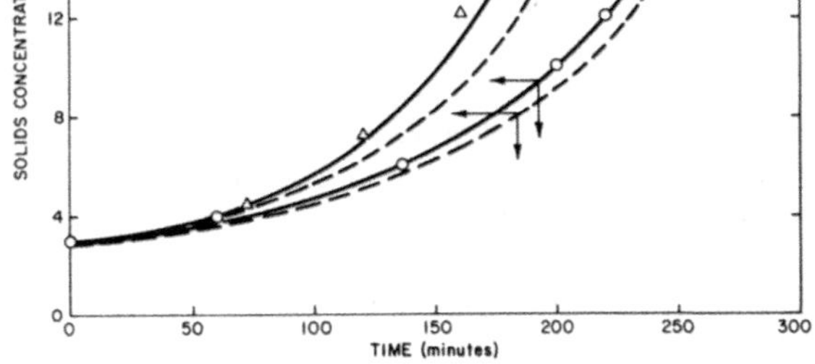

Figure 24. Time required to concentrate a high-protein meal by batch ultrafiltration.

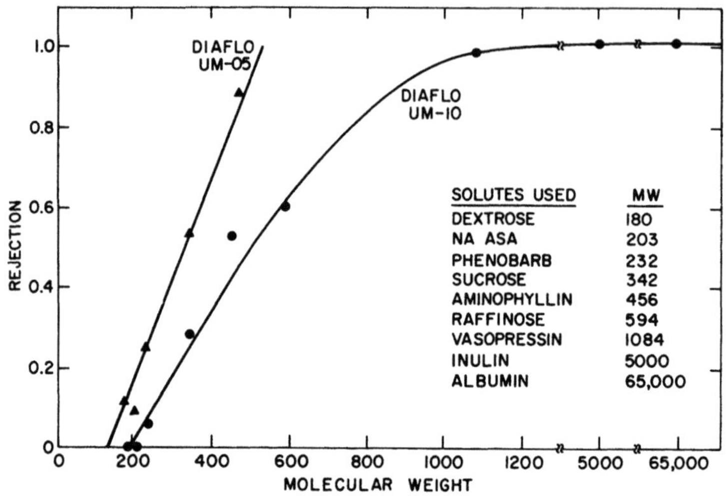

Figure 25. Ultrafiltration selectivities of Diaflo® membranes for biological compounds.

Data were taken in a Model 50 stirred batch cell.

An example of the flux reduction encountered when protein solutions are ultrafiltered was shown in Figure 3 (6.5 percent albumin concentration is approximately equivalent to the concentration of protein in blood plasma). The effects of flow parameters on the magnitude of the flux reduction were extensively studied (in cooperation with the Hospital of the University of Pennsylvania under a contract from the National Institute of Arthritis and Metabolic Diseases) using flat sheet membranes and thin, triangular blood channels. Typical results are shown in Figures 26 through 31. As predicted by Equation (26), the ultrafiltration flux varies as about the cube root of the fluid shear rate at the membrane surface (Figure 26) and inversely as the cube root of the channel length (Figures 27 and 28). The predicted inverse logarithmic dependence of flux on concentration was shown with plasma (Figure 29) with a "gel" concentration of about 58 percent (8.9 times that of whole plasma). As shown in Figure 30, the actual values of the ultrafiltration flux could be made to agree with values predicted from Equation (25) by assuming a value of the average retained protein diffusivity of $D_s = 0.6 \times 10^{-7}$ cm^2/sec. Fluxes were essentially independent of time from the first few minutes of running up to over 30 hours (Figure 31).

Figures 5, 6, and 32 provide illustrations of the independence of flux on pressure and pure-water flux of the membrane, provided that both the pure-water rate is significantly higher than the plasma flux and the pressure is high enough to ensure operation in the gel-polarized region. These data confirm the hypothesis that the fluxes from plasma and blood under the operating conditions used were controlled by back transport from the polarized protein layer at the membrane surface.

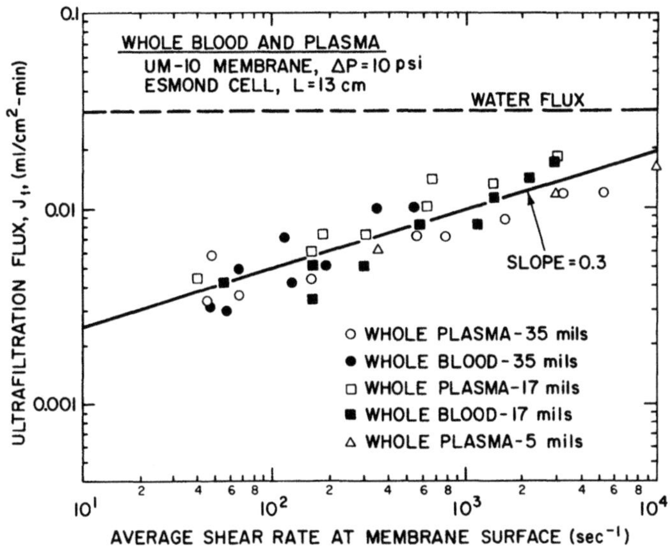

Figure 26. Ultrafiltration of whole blood and blood plasma in a thin-channel laminar-flow recirculating system.

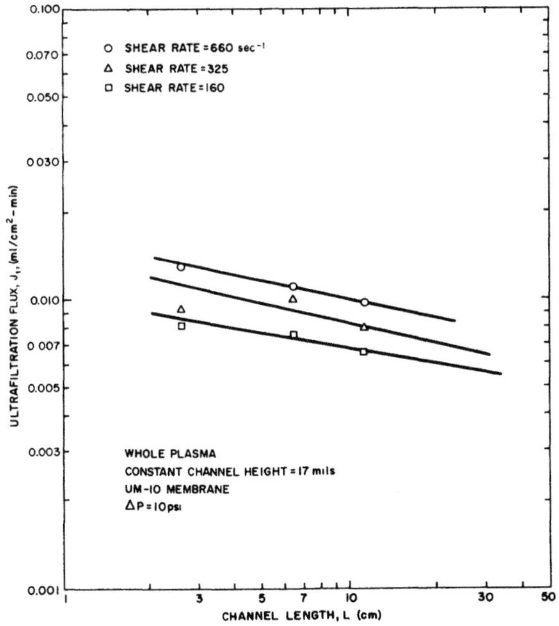

Figure 27. Ultrafiltration flux for human blood plasma in a thin-channel laminar-flow recirculating system – effect of channel length.

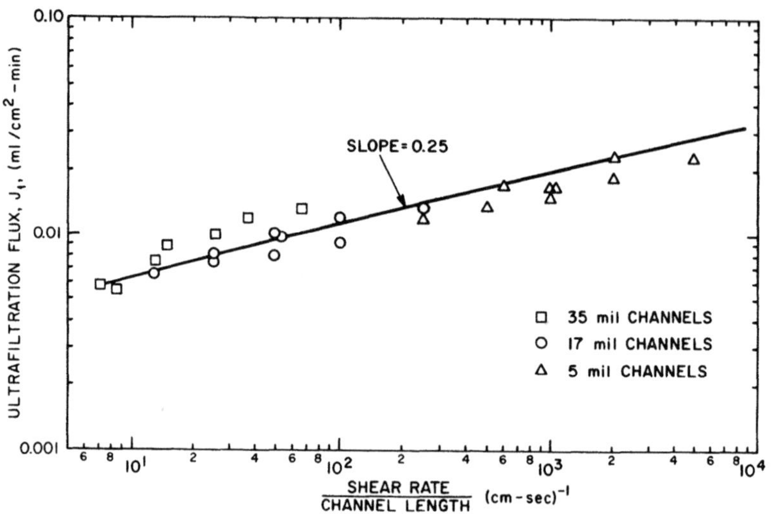

Figure 28. Ultrafiltration flux from human blood plasma in a thin-channel laminar-flow recirculating system – effect of wall shear rate and channel length.

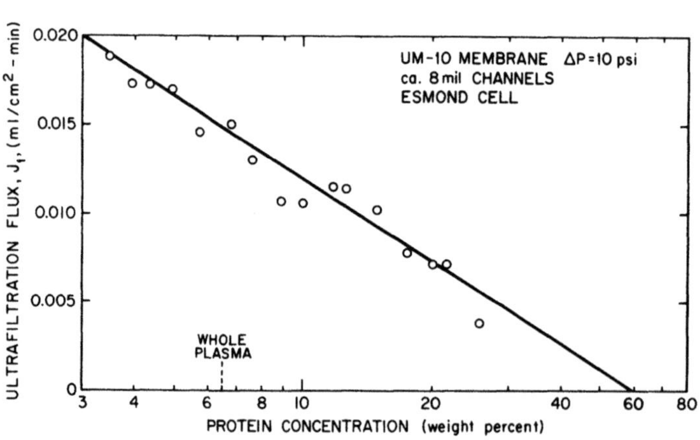

Figure 29. Effect of concentration on the ultrafiltration rates of human blood plasma in a thin-channel laminar-flow recirculating system.

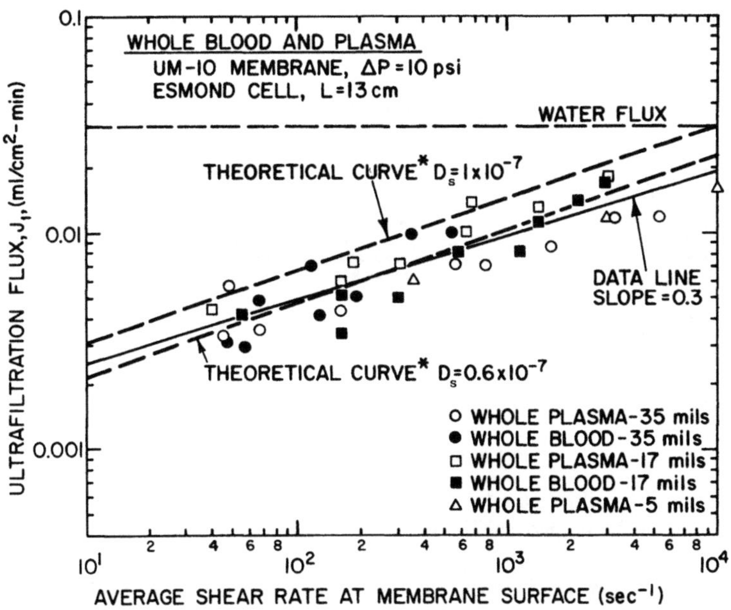

Figure 30. Correlation of ultrafiltration rates from human blood and blood plasma with the solution for mass transfer from a wall of constant concentration. Theoretical curves based on the equation $J_1 = 49 \ln C_B/C_g\, [\dot{\gamma}_w D_s^2/L]^{1/3}$. Data and correlations are for a thin-channel laminar-flow system.

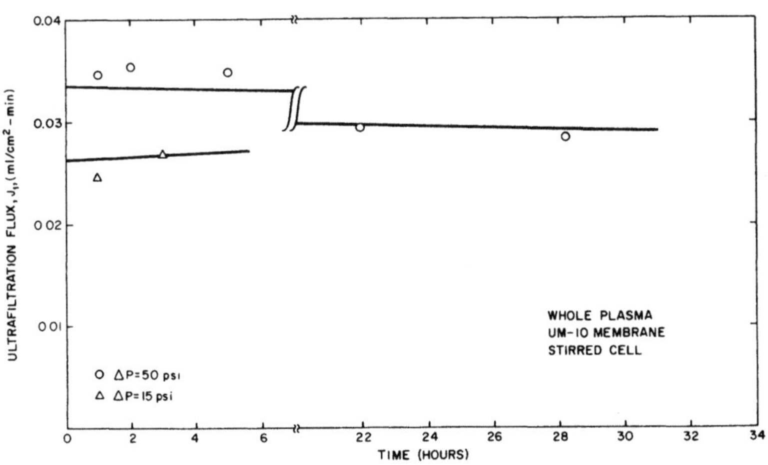

Figure 31. Invariability of ultrafiltration rates with time for human blood plasma in a stirred batch cell.

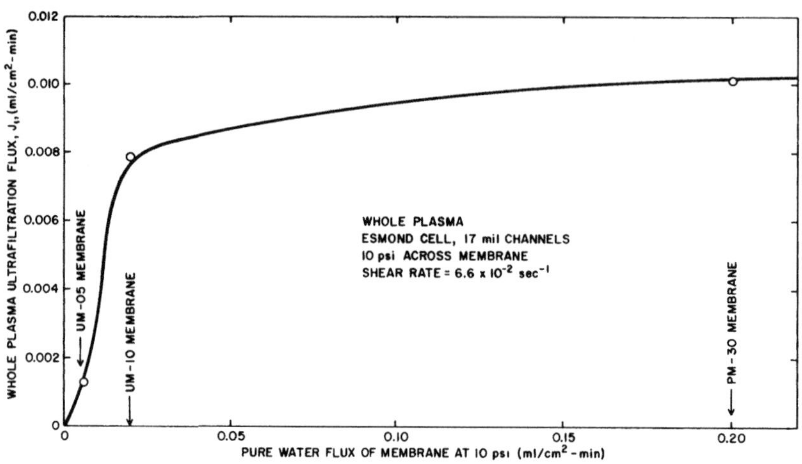

Figure 32. Ultrafiltration rates from human blood plasma through membranes of differing permeabilities to pure water.

Data were obtained in a thin-channel laminar-flow recirculating system.

Concentration and Purification of Albumin Solutions by Ultrafiltration

The intent of this study was to examine the changes in ultrafiltrate flux from bovine-serum albumin solutions as a function of (1) pressure, (2) solute concentration, and (3) recirculation rate.

Operation was conducted under constant concentration conditions, obtained by reconstituting simultaneous with ultrafiltration. A thin channel unit equipped with a PM-10 membrane and a four-port (16-in. path length, 1/4-in. width) stainless steel spacers of 10-mil channel height was used for the ultrafiltration studies.

The variables used were

(1) Bovine serum albumin: 0.1, 1, 5, and 10 percent concentrations by weight, prepared in 0.9 percent NaCl

(2) Pressure: 10, 20, and 40 psi

(3) Recirculation rate: 200, 500, and 1,000 ml/min.

Data are presented in Table 7 and in Figures 33 through 35 and in Figures 12, 13, and 14 presented earlier.

TABLE 7. THIN CHANNEL STUDY (PERISTALTIC PUMP)

Diafiltration of bovine serum albumin solutions at varying pressures and recirculation rates (short-term observations).[a]

Pressure	Recirculation Rate, cc/min	$\frac{2\dot{\gamma}_\omega}{L}$ (cm-sec)$^{-1}$	Flux, gfd 0.1% Alb.	1.0% Alb.	5% Alb.	10% Alb.
10	200	600	26.7	15.6	9.5	9.5
10	500	1,500	33.5	19.6	15.2	13.7
20	200	600	35.4	24.0	19.0	14.7
20	500	1,500	33.1	25.5	19.0	19.0
20	1,000	3,000		35.0	25.9	22.1
40	200	600	55.3	38.1	26.6	20.6
40	500	1,500	55.3	40.0	32.4	26.3
40	1,000	3,000	57.2	43.1	34.3	30.9

(a) Membrane: DIAFLO PM-10; Water Flux: 23 gfd @ 10 psi.

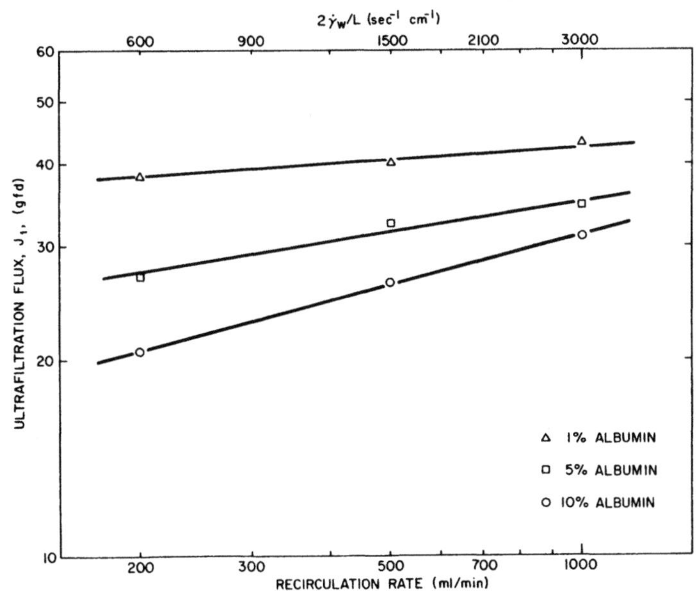

Figure 33. Ultrafiltration of bovine-serum albumin in a TC-1 thin-channel laminar-flow recirculating system at 40 psi and 25 C, using a Diaflo® PM-30 membrane.

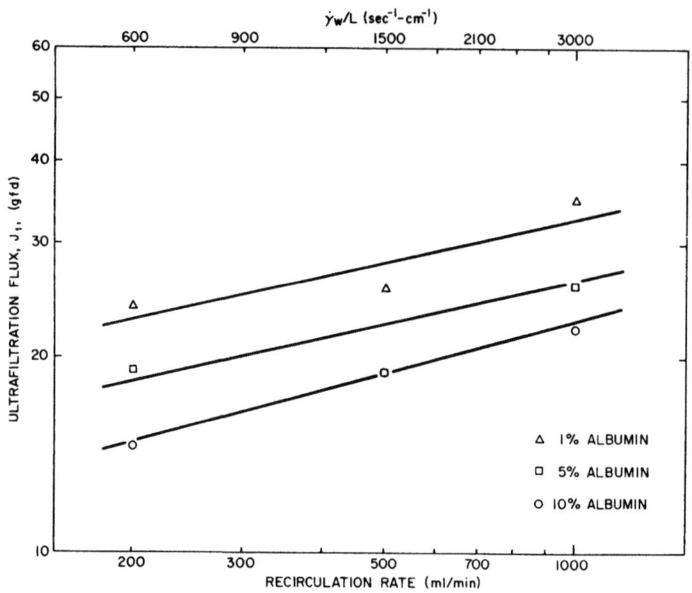

Figure 34. Ultrafiltration of bovine-serum albumin in a TC-1 thin-channel laminar-flow recirculating system at 20 psi and 25 C, using a Diaflo® PM-30 membrane.

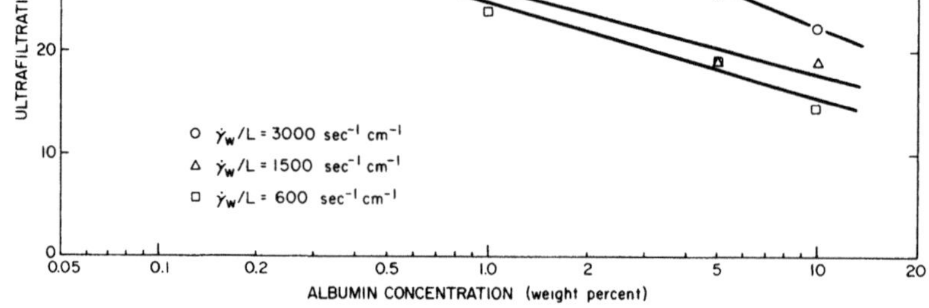

Figure 35. Effect of concentration on the ultrafiltration rates from bovine-serum solutions. Data were obtained in a TC-1 thin-channel laminar-flow recirculating system at 20 psi and 25 C, using a Diaflo® PM-30 membrane.

Pressure Dependence. Figures 12, 13, and 14 presented earlier showed significant increases in ultrafiltration fluxes with increasing pressure, although the fluxes do appear to be "leveling off" as the pressure is increased above about 50 psi. This effect is discussed more fully in the section "Concentration Polarization in the Ultrafiltration of Macromolecules in Solution" in terms of operation below the gel-polarized region.

Effect of Concentration on Flux. Figure 35 shows an approximately logarithmic dependence of the ultrafiltration flux on solution concentration at any given pressure or fluid shear rate at the membrane surface.

The variability in the intercepts of the curves (i.e., in the apparent "C_g") for the various runs and the intercepts above 100 percent concentration are probably also indicative of operation below the gel-polarized region.

PRINCIPLES OF ULTRAFILTRATION DEVICE AND SYSTEM DESIGN

Because of extremely low diffusivities of macromolecules and colloidal solids in suspension, the minimization of concentration polarization (and realization of high ultrafiltration rates) is far more critical for such solutions or suspensions than for the solutions of microsolutes. Concentration polarization is minimized by maximizing the mass-transfer coefficient between the membrane and the fluid for a given set of external operating conditions such as the volumetric feed rate and the required concentration change. For a given volumetric flow rate, the mass-transfer coefficient can be maximized by maximizing the wall shear rate. In turbulent-flow systems, this can be achieved by having large recirculation rates that involve large pressure drops. In the laminar-flow systems, high mass-transfer coefficients can be achieved with little or no recirculation without appreciable pressure drops. This conclusion follows from the fact that the ratio of mass-transfer coefficient per unit pressure drop in a turbulent-flow channel to that in a laminar-flow channel is given by

$$\frac{(k_s/\Delta P)_{turb}}{(k_s/\Delta P)_{lam}} \cong 2.5\, N_{Re_L}^{-0.3}\, (1+R)^{-0.95} \left(\frac{L}{d_h}\right)^{0.33} \qquad (43)^*$$

where N_{Re_L} is the Reynolds number in a laminar channel without recirculation and where R is the recirculation rate required per unit of feed rate in order to make the same channel turbulent.

The corresponding ratios for mass-transfer coefficient and pressure drop are

$$\frac{(k_s)_{turb}}{(k_s)_{lam}} = 0.0125\, N_{Re_L}^{0.45}\, (1+R)^{0.8} \left(\frac{L}{d_h}\right)^{0.33} \qquad (44)^*$$

$$\frac{(\Delta P)_{turb}}{(\Delta P)_{lam}} = 0.005\, N_{Re_L}^{0.75}\, (1+R)^{1.75} \quad . \qquad (45)^*$$

*Derived in the Appendix.

It is clear that one can increase k_S in a turbulent-flow channel by increasing R but only at the cost of more than proportionate increase in the pressure drop across the channel. With a minimum overall processing cost as the basis for design, two cases must be considered, equipment for laboratory or industrial scales.

Laboratory-Scale Equipment

In this case, the flow rates are small enough so that a certain minimum fixed charge must be incurred on the pumping and other equipment. These charges are large compared to the operating cost for these pumps. In such a case, the excessive pressure drop resulting from the use of a turbulent channel can probably be tolerated, while the accompanying increase in k_S gives higher fluxes.

Industrial-Scale Equipment

Where flow rates are high and therefore operating costs are large compared with fixed costs, laminar-flow thin-channel devices (channels) seem to be more economical, as they provide higher mass-transfer coefficients per unit pressure drop for a given flow rate. It has also been pointed out in the earlier analysis that the average mass-transfer coefficient in a laminar-flow channel is related inversely to the channel length. Therefore, operation with a number of short channels in parallel is to be preferred to operation with fewer long channels. There is, of course, some feasibility limitation on the number of short parallel flow paths that can be embodied into a large-area ultrafiltration-membrane module.

It must be pointed out that the above design considerations apply only to systems in which ultrafiltration fluxes are limited by concentration polarization alone. The analysis is therefore valid only under operating conditions where ultrafiltration flux is pressure independent and is, thus, solely controlled by gel polarization.

Apart from the above analysis, some other design considerations are rather obvious. An ultrafiltration device must provide proper mechanical support for the membrane, uniform distribution of the process liquid over the membrane surface, and adequate flow rates for the passage of feed, retentate, and ultrafiltrate with minimum pressure drop; it must have a high active membrane-area/volume ratio, must be simple and easy to clean out and to replace membranes and reassemble, must be reliable as regards leakage (particularly between feed and ultrafiltrate compartments) and dimensional changes under pressure. Finally, the device must be inexpensive to fabricate, repair, and maintain. Accommodating all these requirements into a practicable and economic membrane-ultrafiltration module is the major challenge confronting the designer and fabricator of reliable and efficient ultrafiltration equipment and systems.

SUMMARY AND CONCLUSIONS

The development of very high-flux ultrafiltration membranes for the separation and concentration of micromolecular, macromolecular, and colloidal solutions has brought to prominence a problem that was usually of negligible influence with older low-flux membranes: that of concentration polarization of solutes at the upstream membrane surface. The concentrated layer thus formed invariably exerts a deleterious influence on solvent flux and solute separation efficiency, so that the understanding and control of this phenomenon becomes of major importance. Until recently, however, only microsolute polarization has received much study.

In this paper, microsolute polarization has been reviewed, and models have been presented for the concentration polarization of macrosolutes and colloids along with experimental data supporting these models. For all these solutions, the resulting solvent fluxes have been shown to be controlled by the back-transport rate of concentrated solute from the upstream membrane surface into the bulk solution above the membrane. Achievement of high fluxes depends, therefore, upon operating at flow and concentration conditions that maximize the rate of mass transfer from the membrane surface. In laminar-flow systems, this is achieved by operating at high fluid velocities across the membrane surface in thin, short channels. Increasing the concentration of the solution decreases the flux. In turbulent-flow systems, fluxes are maximized by operation at high Reynold's numbers in thin channels with dilute solutions.

Solute-separation efficiency has also been shown to be complexly dependent on the degree of concentration polarization and on the specific solutes making up the polarized layer. Quantitative modeling of these effects for macrosolutes and colloids has not yet been achieved.

These analyses present practical starting guidelines for the design of efficient ultrafiltration systems. Translation of these guidelines into building of practical devices and systems for ultrafiltration continues to provide a challenge to the engineer who wishes to make economical use of this new and attractive separation process.

ACKNOWLEDGMENTS

The authors would like to acknowledge the substantial insights on data analysis provided by discussions with Dr. Robert W. Hausslein and Dr. Harris J. Bixler of the Amicon Corporation. Significant contributions toward our understanding of concentration polarization in ultrafiltration systems were also provided by Professor C. Judson King of the University of California at Berkeley.

REFERENCES

(1) Vink, Hans, *Acta Chem Scand.,* **20** (8), 2245-9 (1966).

(2) Johnson, J. S., Dresner, L., and Kraus, K., "Hyperfiltration (Reverse Osmosis)", Chapter 8 in *Principles of Desalination,* K. S. Spiegler, Ed., Academic Press, New York (1966).

(3) Spiegler, K. S., and Kedem, O., *Desalination* (1), 311-326 (1966).

(4) Kedem, O., and Katchalsky, A., *J. Gen. Physiology,* **45**, 143 (1961).

(5) Brian, P.L.T., *I&EC Fund,* **4**, 439 (1965).

(6) Sherwood, T. K., Brian, P.L.T., Fisher, R. E., and Dresner, L., *I&EC Fund,* **4**, 113 (1965).

(7) Gill, W. N., Tien, C., and Zeh, D., *I&EC Fund,* **5**, 367 (1966).

(8) Kimura, S., and Sourirajan, S., *I&EC Proc. Design and Dev.,* **7** (1), 41-48 (1968).

(9) Chilton, T. H., and Colburn, A., *Ind. Eng. Chem.,* **26**, 1183 (1934).

(10) Blasius, H., *Z. Math. Phys.,* **56**, 1-37 (1908).

(11) Fisher, R., Ph.D. Thesis, MIT, Department of Chemical Engineering (1961).

(12) Dresner, L., Oak Ridge National Lab *3621* (May 1964).

(13) Shor, A. J., Kraus, K. A., Johnson, J. S., and Smith, W. T., *I&EC Fund,* **7** (1), 44-48 (1968).

(14) Johnson, Kraus, et al., French Patent 1,497,295.

(15) Graetz, L., *Ann. d. Physik.,* **25**, 337-357 (1885).

(16) Leveque, J., *Ann. Mines,* **13**, 201, 305, 381 (1928).

(17) Calderbank, P. H., and Moo-Young, M. B., *Chem. Eng. Sci.,* **16**, 34 (1961).

(18) Chilton, T. H., and Colburn, A., *Ind. Eng. Chem.,* **26**, 1183 (1934).

(19) Colton, C. K., Ph.D. Thesis, MIT, Dept. of Chem. Eng. (1969).

(20) Marangozis, J., and Johnson, A. I., *Can. J. Chem. Eng.,* **40**, 231 (1962).

(21) Baker, R. W., *J. Appl. Poly. Sci.,* **13**, 369-376 (1966).

(22) Weissman, B. J., Smith, C. V., Jr., and Okey, R. W., *Chem. Eng. Prog. Symp. Series,* **64** (90), 285 (1968).

(23) Work performed by the Amicon Corporation under support from the Fats and Proteins Research Foundation, Inc., of Des Plaines, Illinois.

(24) Bixler, H. J., Nelsen, L. M., and Besarab, A., "The Diaphron Hemodiafilter: An Alternative to Dialysis for Extracorporeal Blood Purification", *Chem. Eng. Prog. Symp. Series,* **64** (84), 90-103 (1968).

(25) Bixler, H. J., Nelsen, L. M., and Bluemle, L. W., Jr., *Trans. Amer. Soc. Artif. Int. Organs,* **XIV**, 99-108 (1968).

(26) Henderson, L. W., Besarab, A., Michaels, A. S., and Bluemle, L. W., Jr., *Trans. Amer. Soc. Artif. Int. Organs,* **XIII**, 216 (1967).

(27) Most of the work reported here was conducted by Amicon Corporation in cooperation with the Hospital of the University of Pennsylvania under support from the National Institute of Arthritis and Metabolic Diseases, Contract No. PH43-66-45.

NOMENCLATURE

A_m	Membrane area
B	Constant in the Leveque solution [see Equation (26)]
C_B	Solute concentration in the bulk solution above the membrane
C_F	Final concentration of retentate
C_f	Filtrate concentration
C_g	Gel concentration
C_o	Initial solute concentration
C_w	Solute concentration at the upstream membrane surface (wall concentration)
c	Solute concentration at a given point
D_s	Solute diffusion coefficient
d	Diameter of packed particles [Equation (24)]
d_h	Equivalent diameter
f	Fanning friction factor
g_c	Gravitational acceleration

h Half channel height

J_1 Solvent flux

J_1^o Solvent flux in the absence of polarization

J_2 Solute flux

j_D Chilton-Colburn factor

k_s Mass-transfer coefficient for solute transport away from the membrane surface

L Channel length

M Polarization modulus ($M = C_w/C_B$)

N_{Re} Reynolds number $\left(N_{Re} = \frac{d_h u_B \rho}{\eta}\right)$

N_{Re_L} Reynolds number in the laminar-flow regime

N_{Sc} Schmidt number ($N_{Sc} = \nu/D_s$)

P Permeability of porous solid

$\bar{P}$ Average boundary-layer permeability

P_g Permeability of a uniform gel layer of concentration C_g

$\bar{P}_o$ Average boundary-layer permeability when $C_w = C_g$

$\bar{P}_1$ Membrane permeability to solvent

$\bar{P}_2$ Membrane permeability to solute

Δp Pressure drop in the direction of bulk flow

ΔP Applied pressure drop across membrane

Q Volumetric flow rate in the channel

R Recirculation rate per unit feed rate in a turbulent channel

R_g Flux resistance of the gel layer

R_m Flux resistance of the membrane

R_p Resistance of polarized layer

r Cell radius in stirred cell

t_b	Thickness of boundary layer ($t_b = \delta$)
t_g	Thickness of concentration boundary layer
t_m	Effective thickness of membrane
u_B	Bulk stream velocity
$u_{(o)}$	Velocity of fluid entering the channel
V	Volume eluted from chromatography column (Table 2)
V_o	Void volume of column (Table 2)
V_o	Starting volume
V_w	Wall permeation velocity

GREEK SYMBOLS

$\dot{\gamma}_\omega$	Fluid shear rate at the membrane surface
δ	Concentration boundary-layer thickness
ϵ	Porosity of solid (void fraction)
η	Solvent viscosity
θ	Time
ν	Kinematic viscosity ($\nu = \eta/\rho$)
ξ	$\left[= \dfrac{V_w^3\, h\, L}{3\, u_{(o)}\, D_s^2}\text{; see Equation (19)}\right]$
$\Delta\pi$	Osmotic-pressure difference across membrane
ρ	Fluid density
σ	Rejection coefficient
σ_A	Apparent rejection coefficient [see Equation (7a)]
ω	Stirrer speed

APPENDIX

Comparison of Mass Transfer in Laminar and Turbulent Channels

Let

Q = Volumetric flow rate in laminar channel

R = Recirculation per unit main flow in turbulent channel

N_{Re} = Reynolds number in turbulent channel

N_{Re_L} = Reynolds number in laminar channel

$$\Delta p_L = \frac{32}{N_{Re_L}} \frac{L\,Q^2\,\rho}{\frac{\pi}{4}\,g_c d_h^{\,3}}$$

$$\Delta p_T = \frac{2 \times 0.079\ Re_T^{-0.25}\ L\,Q^2\,\rho}{\frac{\pi}{4}\,g_c d_h^{\,3}}$$

But

$$N_{Re_T} = (1 + R)\,N_{Re_L}$$

$$\therefore \Delta p_T \cong \frac{0.16\,(1 + R)^{-0.25}\,N_{Re_L}^{-0.25}\ L\,Q^2\,\rho\,(1 + R)^2}{\left(\frac{\pi}{4}\right)\,g_c d_h^{\,3}}$$

$$\therefore \frac{\Delta p_T}{\Delta p_L} = \frac{0.16\,(1 + R)^{1.75}\,N_{Re_L}^{-0.25}}{\left(\frac{32}{N_{Re_L}}\right)}$$

$$= \frac{1}{200}\,(1 + R)^{1.75}\,N_{Re_L}^{\;0.75}$$

Also

$$\frac{k_{sL} d_h}{D} = 1.86\,N_{Re_L}^{\;0.33}\,N_{Sc}^{\;0.33}\left(\frac{d_h}{L}\right)^{0.33}$$

$$\frac{k_{sT}d_h}{D} = 0.023\, N_{Re_T}{}^{0.8}\, N_{Sc}{}^{0.33}$$

$$\therefore \frac{k_{sT}}{k_{sL}} = \frac{1}{80}\, N_{Re_L}{}^{0.8}\, (1 + R)^{0.8}\, N_{Re_L}{}^{-0.33} \left(\frac{L}{d_h}\right)^{0.33}$$

$$= \frac{1}{80} N_{Re_L}{}^{0.45}\, (1 + R)^{0.8} \left(\frac{L}{d_h}\right)^{0.33}$$

$$\therefore \frac{(k_s/\Delta p)_{Turb}}{(k_s/\Delta p)_{Lam}} = 2.5\, N_{Re_L}{}^{-0.3}\, (1 + R)^{-0.95} \left(\frac{L}{d_h}\right)^{0.33}$$

Similarly

$$\frac{(k_s)_{Turb}}{(k_s)_{Lam}} = 0.0125\, N_{Re_L}{}^{0.45}\, (1 + R)^{0.8} \left(\frac{L}{d_h}\right)^{0.33}$$

and

$$\frac{(\Delta p)_{Turb}}{(\Delta p)_{Lam}} = 0.005\, N_{Re_L}{}^{+0.75}\, (1 + R)^{1.75}$$

ENZYME PROCESSING USING ULTRAFILTRATION MEMBRANES

Daniel I. C. Wang, Anthony J. Sinskey, and Thomas A. Butterworth
Department of Nutrition and Food Science
Massachusetts Institute of Technology
Cambridge, Massachusetts 02139

In the past several years there has been a continual increase in the use of ultrafiltration (UF) membranes for concentrating, separating, and purifying various dissolved solutes. The forecast in membrane separation processes as reported by Pattison (1968) will amount to $75 million by 1975. Of this total, it was estimated that $37.5 million will be spent in UF processes for industrial, medical, and surgical operations. Parallel in time with membrane-technology development is the increased usage of various enzyme preparations. Specifically, the use of enzymes in laundry detergents has caught the fancy of housewives in the past 2 years. Although the enzyme industry in the United States is by no means new, the recent innovation of enzyme additives in detergents has contributed significantly to the total market. For example, the total U. S. enzyme market in 1968 amounted to approximately $40 million, of which approximately $9 million was for laundry products (Anon., 1969A). However, the latter is expected to increase to $25 to $30 million in 1969 (Anon., 1969B).

The detergent enzymes are all produced by microorganisms in submerged fermentations. Having achieved a substantial enzyme market, one is, therefore, constantly searching for methods which could improve the methods of recovery. Enzymes, being macromolecular proteinous substances produced at low concentrations during fermentation, appear to be a natural for the employment of ultrafiltration membranes for their recovery. In addition to the detergent enzymes, there is every reason to believe that membrane processing could contribute to the recovery, separation, and purification of other industrial and research enzymes already on the market.

Before proceeding to outline the objectives and findings on the use of ultrafiltration membranes for enzyme processing, it would be desirable to present a brief summary of this subject and also allied subjects which have already appeared in the literature. The general subject of ultrafiltration dealing with the theory and application has been excellently documented by Michaels (1968A). The industrial applications of ultrafiltration membranes was, however, presented only in a qualitative form. More recently, Michaels (1968B) presented some additional information on the performances of various types of ultrafiltration membranes. Some basic economics on UF processing along with the design configurations of large-scale processing units were also presented. The reader is encouraged to review these documents as well as the book by Rickles (1967) dealing with ultrafiltration-membrane technology and economics.

The uses of UF membranes in biological recovery processes at the laboratory scale are beginning to appear in the literature. One of the earliest reports on the use of UF membranes for concentrating biological materials (proteins) was that presented by Blatt et al. (1965). These authors pointed out that large increases in flux can be attained for UF membranes as compared with the conventional cellulose acetate dialysis membranes. Since then, many publications have appeared dealing with the concentration, separation, and purification of proteins and enzymes. For example, Wang et al. (1968) have successfully concentrated several enzymes and a bacteriophage by the use of UF membranes. Protein separations by these membranes have also been successfully performed, as illustrated by Zipilivan et al. (1969). These investigators separated immunologically active materials by means of membrane-partition techniques. In addition, Blatt et al. (1967A) fractionated a mixture of proteins and smaller organic compounds by membrane chromatography.

If a membrane separation technique is successful, purification will generally be accomplished. These types of results have also been reported by Blatt et al. (1967B) for the protein bovine α-lactalbumin and by Penhoet et al. (1967) for the enzyme fructose diphosphate aldolose. It would be beyond the scope of this paper to review exhaustively all of the documents in UF processing. It was, however, the intent of the authors to present a brief summary as to some of the available literature pertinent to this paper.

The primary objective of this paper is to present the findings from our laboratory in the general area of enzyme processing using ultrafiltration membranes. Some of the results were accumulated in our laboratory dating back to 1966, while others were completed only very recently. The presentation in this paper will be subdivided into three general sections. The first of these will deal with the concentration and purification of different enzyme preparations by UF membranes. This will be followed by the findings on the production of an extracellular enzyme in a "membrane fermentor". The last section will deal with the continuous enzymic reactions in a "membrane reactor".

MATERIALS AND METHODS

Enzyme Concentration and Purification

Ultrafiltration Apparatus and Procedure. The apparatus used for the ultrafiltration studies is shown schematically in Figure 1 (Amicon Corp., Lexington, Mass.). For filtering small volumes, solutions up to 450 ml are placed directly in the polycarbonate cylinder (5). For handling volumes greater than 450 ml, the solution is placed in a fiberglass feed reservoir (7) which has a total capacity of about 12 liters. The hydraulic pressures for ultrafiltration were maintained by means of high-pressure cylinder nitrogen. This filtration unit is capable of withstanding pressures up to 100 psig.

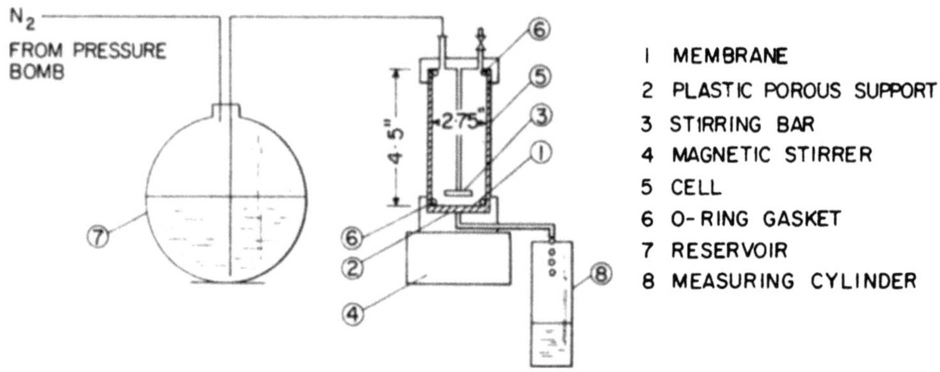

Figure 1. Schematic diagram of ultrafiltration apparatus.

The filtration cell is made of two stainless end caps mounted on the top and bottom of the polycarbonate cylinder. The membrane, 2-7/8 inch in diameter, is mounted on a porous plastic disc (2). Leakage is prevented by means of two O-rings (6) mounted directly above and below the cell. The top end cap contained several ports for the introduction of feed and the removal of samples. A magnetic stirring bar (3), located approximately 1/8 inch above the membrane, is constantly rotated by means of a stirrer (4) to minimize product buildup at the membrane surface. The filtrate passes through the bottom end cap and is collected in a measuring cylinder. The entire cell unit can be immersed in a constant-temperature bath for conducting studies at various temperatures.

The membranes which were employed in this study were supplied by two manufacturers. The UM-1 membranes (Amicon Corp., Lexington, Mass.) are reported by the manufacturers to be capable of retaining a nominal molecular weight of 10,000. The HFA-200 membranes (Abcor, Inc., Cambridge, Mass.) are reported capable of retaining macromolecules of 20,000 molecular weight.

Enzyme Assays. Three different enzymes were studied in the use of UF membranes for their processing. The source of the enzyme along with the assay technique employed are now briefly outlined.

(1) *Penicillinase Assay.* Purified penicillinase was obtained through the courtesy of Yissum Research and Development Company, Hebrew University, Jerusalem, Israel. The enzyme was suspended in 0.1 M, pH 7.0, phosphate buffer (KH_2PO_4 and Na_2HPO_4). The assay procedure of Citri (1964) was used for determining the penicillinase activity. An enzyme unit for penicillinase is defined as the amount of enzyme that forms 1 micromole of penicilloic acid in 1 hr at 30°C. The specific activity of the enzyme preparation was 70,000 units/mg protein.

(2) *β-galactosidase Assay.* Yeast β-galactosidase (Grade B, Calbiochem, Los Angeles, Calif.) was suspended in 0.1 M, pH 8.2, Na_2HPO_4 containing 0.001 M $MgSO_4$, 0.002 M $MnSO_4$ and 0.1 M β-mercaptolthanol. The assay reported by Schlomo et al. (1955) was used to determine

the enzyme activity. The enzyme unit is defined as the amount of enzyme that will catalyze the hydrolysis of 0.001 M o-nitrophenylgalactoside at pH 7.5 at 30°C in 0.07 M Na_2HPO_4 at the rate of 0.012 micromole per ml of reaction mixture. The specific enzyme activity was approximately 13 units/mg sample.

(3) *Trypsin Assay.* Trypsin (1-300 Nutritional Biochemicals Corp., Cleveland, Ohio) was suspended in 0.1 M, pH 8.2, Tris buffer (Tris hydroxylmethyl aminomethane) containing 0.02 M $CaCl_2$. The assay procedure of Garnot (1966) was used to determine the trypsin activity. One enzyme unit is defined as the amount of enzyme that forms 1 micromole of p-nitroaniline per ml of reaction mixture per minute at 25°C. The activity of the enzyme was found to be 45 units/mg protein.

(4) *Protein Assay.* Protein was analyzed according to the method of Lowry et al. (1951).

Sephadex Gel Liquid Chromatography. Gel-filtration studies were performed on various enzyme (trypsin) samples from the ultrafiltration experiments. Sephadex, G-75 gel (Pharmacia Fine Chemicals, Inc., Piscataway, N. J.), reported to be capable of fractionating molecular weights ranging from 1,000 to 50,000, was used in a 15-mm-diameter, 500-mm-high column. Five ml of the sample was placed onto the column and eluted with 0.1 M, pH 8.2, Tris buffer containing 0.02 M $CaCl_2$. Elution rate through the column was maintained at 40 ml/hr. Samples of the eluate were collected in 3-ml fractions and measured quantitatively for their respective trypsin and protein concentrations.

Molecular-weight markings of the Sephadex gel were performed using 3X crystalline ribonuclease (Code R, Worthington Biochemical Corp., Freehold, N. J.), 2X crystalline trypsin (Bovine, Gallard-Schlesinger Chemical Corp., L. I., N. Y.), and blue dextran 2000 (Pharmacia Fine Chemicals, Inc., Piscataway, N. J.). The molecular weight of ribonuclease was reported by Hirs et al. (1956) to be 13,683. The molecular weight of trypsin was reported by Kay et al. (1961) to be 23,800. The molecular weight of blue dextran 2000 was reported by the manufacturer to be 2,000,000.

The Membrane Fermentor: Methods and Procedures

This section of this paper deals with the coupling of an UF membrane with a fermentor in order to examine this influence on the production of extracellular proteolytic enzymes.

Apparatus. The apparatus which was used in these studies was essentially that shown previously in the schematic diagram of Figure 1. The only differences were that sterile and aseptic techniques had to be employed. In these experiments, the cell (5) in Figure 1 played a dual role, acting as a fermentor as well as a filtration cell. Sterile medium was placed in the fiberglass reservoir (7) and, again, sterile cylinder nitrogen was employed for filtration pressure. The filtration cell and the connecting lines were sterilized with steam at 125°C, 15 psig, for 15 minutes. The membrane, porous plastic membrane support, and the feed reservoir were sterilized with 70 percent ethanol, followed by washing with sterilized distilled water. The unit was carefully assembled to insure sterility.

The membrane which was employed in these studies was HFA-300 (Abcor, Inc., Cambridge, Mass.). This membrane was selected since its molecular-weight cutoff was reported by the manufacturer to be on the order of 55,000. The proteolytic enzymes which are excreted by the microorganisms had been speculated by Keller and Mandl (1963) to be on the order 100,000 in molecular weight.

Membrane Fermentation. The fermentation selected for this membrane system was the production of extracellular proteases by *Clostridium histolyticum* (ATCC 8034). The organism was employed in view of its ability to produce large amounts of proteases extracellularly and anaerobically. The latter condition would enhance the ease of maintaining asepsis as well as eliminating oxygen-transfer requirements during fermentation. It should be pointed out, however, these are not necessary criteria for the successful operations of a membrane fermentor. In our case, it was done for the sake of convenience.

Inoculum for the fermentation was prepared in 500-ml Erlenmeyer flasks using a germination medium and incubated anaerobically for 15 hours at 37°C. At this time, 30 ml of the seed culture was used to inoculate 1000 ml of medium in an anaerobic jar containing the growth medium. Composition for the media are shown in Table 1. Growth in the anaerobic jar was followed by optical density measurements. At the time when the fermentation approached its stationary phase, the culture was divided into two fractions. Approximately 460 ml of the suspension was aseptically removed and placed into the membrane fermentor. The remainder was allowed to continue incubation and served as a control sample. At this time, feed to the membrane fermentor was initiated and, simultaneously, filtrate was continuously removed. The temperature during fermentation was always maintained at 37°C.

TABLE 1. MEDIA COMPOSITION FOR PROTEASE PRODUCTION BY *Clostridium histolyticum*

A. Germination Medium	
Yeast extract (Difco)	1.0 gm
Protease-peptone (Difco)	2.5 gm
Na_2HPO_4	0.45 gm
KH_2PO_4	0.1 gm
$MgSO_4 \cdot 7H_2O$	0.05 gm
$FeSO_4 \cdot 7H_2O$	0.05 gm
Distilled water	100 ml
pH before sterilization	7.6
B. Fermentation Medium	
Protease-peptone (Difco)	5.0 gm
Casamino acids (Difco)	2.0 gm
Distilled water	100 ml
pH before sterilization	7.5

During the membrane fermentation, the following analyses were performed: cell dry weight, protease activity, pH, and ultrafiltration rate. The cell dry weights were obtained by centrifuging a sample, followed by two distilled-water washings and drying at 100°C for 24 hours. The protease assay of the broth had to be modified from the conventional casein assay due to the large amount of potentially interfering substances (amino acids) in the fermentation medium. The essential procedure of Charney and Tomorelli (1947) was employed. The enzyme substrate, azo casein, was prepared according to the methods of Charney and Tomarelli (1947) and Berman et al. (1961). Using this substrate, it was possible to assay optically the products of proteolysis at 425 μm without interference from the amino acids present in the fermentation broth. Lastly, pH and filtration rate were monitored during the course of the fermentation.

The Membrane Reactor: Methods and Procedure

The last section of this paper deals with continuous enzymic reactions coupled with an ultrafiltration membrane. Two enzyme reactions were studied; the methods and procedure will now be briefly described.

Membrane Reactor. The membrane reactor used in these studies was a model 400 filtration cell (Amicon Corp., Lexington, Mass.) with a 12-liter feed reservoir. Modifications were made in order that a constant liquid level and continuous substrate addition could be maintained during operation. A schematic diagram of this membrane reactor system is shown in Figure 2. The enzyme-substrate mixture first was placed into the reactor. Initially, the system was operated batchwise (usually, for several hours). After operation as a batch system, the entire system was pressurized by means of pressurized nitrogen (A) up to the ultrafiltration pressure of 15 psig. Continuous feed and product removal were then started. The feed from the substrate reservoir (B) is initiated by means of a liquid level probe (J) in combination with a positive-displacement feed pump (C). Fresh substrate is first pumped into an "inverted" surge tank (D). After an appropriate volume (20 ml) is collected in the surge tank, the fresh substrate flows by way of the siphon arrangement (K) into the reaction vessel. This feed arrangement eliminated excessive cycling of the pump. The substrate feed can be analyzed through the sample return port (E) and the products in the reaction vessel can be removed from the sample line (I). A magnetic stirring bar (F) located about 1/8 inch above the membrane (G) was used to decrease product buildup on the membrane. The reaction products were obtained continuously through the exit port (H).

The UF membranes used throughout these studies were designated as PM-10 (Amicon Corp., Lexington, Mass.). These membranes are similar in characteristics to the UM-1 membranes which are reported to have a 10,000-molecular-weight cutoff. The membranes were 2-7/8 inch in diameter. The methods of sealing and assembling the filtration cell were similar to those described previously.

Analytical Techniques: Enzymes, Substrates and Products. Two enzyme systems were employed in the studies on the continuous membrane reactor: α-amylase and glucoamylase. The substrate for both studies was waxy-maize starch (American Maize-Products Co.), and it was selected because it is essentially 99 percent pure amylopectin. At this amylopectin content, the

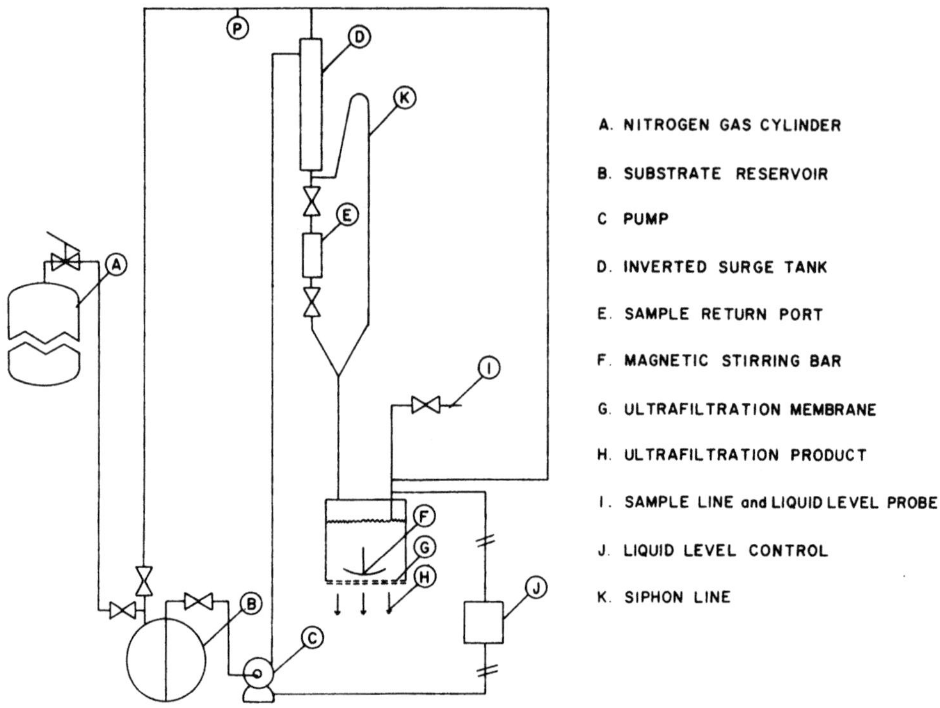

Figure 2. Schematic diagram of membrane reactor.

process of retrogradation (formation of insoluble starch solids) is at a minimum. α-amylase from ***Bacillus subtilis*** (Sigma Chemical Co., Type II-A, St. Louis, Mo.) was selected in view of the large amount of available literature on optimal reaction conditions. In addition, the products of hydrolysis by α-amylase on starch are di-, oligo- and polysaccharides. Elaboration on this selection will be presented in the discussion of the results. The second enzyme, glucoamylase (Grade II, Sigma Chemical Co., St. Louis, Mo.) was selected since the products of reaction are generally glucose monomers. Thus, a comparison of the two enzyme systems should be quite interesting. It should be mentioned, however, that both enzymes have been employed in industrial processes for the saccharification of starch products.

Total carbohydrate concentration in various samples was determined by the anthrone method as outlined by Seifter et al. (1950). The degree of polymerization of a starch molecule was measured as a dextrose equivalent (DE) unit. The procedure used to measure the reducing power of a dextrin mixture was that reported by Ashwell (1957). Dextrose equivalent (DE) is defined as:

$$\mathrm{DE} = \frac{(\text{Reducing Power of Unknown}) \quad (\text{Total Carbohydrate of Std.})\ (100)}{(\text{Reducing Power of Std. Dextrose}) \quad (\text{Total Carbohydrate of Unknown})}$$

Dextrose equivalent of carbohydrates range from zero to 100, with the latter value for pure dextrose.

Glucose was analyzed using the glucose oxidase technique (Glucostat Special, Worthington Biochemical Corp., Freehold, N. J.). This enzyme preparation contains no trace of amylase, maltase, or invertase.

Enzyme activities for both α-amylase and glucoamylase were determined by the reduction of 3,5 dinitrosalicylic acid as outlined by Berfeld (1951). When these enzymes were in the presence of dextrins or other reducing sugars, a modified assay method measuring the rate of decrease of iodine stainability of whole starch, as outlined by Briggs (1967), was employed. The Briggs unit (1967) is defined as 100/t, where t is the time in minutes required for an enzyme solution to reduce the iodine coloring capacity of soluble starch to one-half of its original value under reaction conditions of 25°C, pH 5.6, in 0.02 M $CaCl_2$ solution. The specific activities of α-amylase and glucoamylases were, respectively, 4470 and 3.07 Briggs unit per mg of sample.

Enzyme and Substrate Preparation. In the continuous enzyme studies using α-amylase, the initial enzyme concentration was 0.160 mg/ml. The substrate was prepared by adding a sufficient amount of waxy-maize starch to give a final starch concentrate of 0.4 percent. The starch was first suspended in 5 ml of 0.02 M, pH 5.6, acetate buffer containing 0.002 M $CaCl_2$. This solution was then combined with approximately one-half the final volume of the same buffer solution. This solution was agitated and boiled for about 2 minutes. The remainder volume of buffer was then added and cooled, with a final solution 0.4 percent starch thus resulting.

For the glucoamylase studies, a 9 percent starch solution was prepared in the manner described above. This "paste" was "enzyme thinned" by adding sufficient α-amylase to yield 0.8 microgram of enzyme per ml of solution and allowing partial hydrolysis to occur at 50°C for 10 minutes. A decrease in viscosity and optical density results after the thinning process. This solution, at a dextrose equivalent of 10, is then used as the substrate for continuous glucoamylase reactions.

RESULTS AND DISCUSSION

Enzyme Concentration and Recovery Efficiency

The results of the concentration and recovery efficiencies for various enzyme solutions are summarized in Table 2.

Using the UM-1 membrane, it was possible to achieve a volume concentration factor of 12.3 for penicillinase solution in approximately 8 hours at a filtration pressure of 100 psig. No penicillinase activity was detected in the filtrate. The latter finding is well within reason, since this enzyme is reported to have an average molecular weight ranging from 30,000 to 36,000. However, after UF processing, only 79.3 percent of the initial enzyme activity could be accounted for. A control sample of penicillinase stored in the same buffer system up to 15 hours did not show any appreciable loss in activity. Further analysis as to possible reason for this loss will be presented in a later section.

TABLE 2. CONCENTRATION AND RECOVERY OF ENZYMES USING A UM-1 MEMBRANE

Enzyme	Initial Volume, ml	Final Volume, ml	Volume Concentration Factor	Enzyme Concentration, unit/ml: Original Sample	Concentrate	Filtrate	Recovery Efficiency,[a] percent
Penicillinase	1800	150	12.3	100	950	0	79.3
β-galactosidase	1000	33	30.3	0.75	5.44	0	24[c]
Trypsin[b]	1400	310	4.52	4.5	18.7	0	91.3
Trypsin	1320	290	4.56	4.3	14.9	0	76.6

(a) Recovery efficiency defined as: (total enzyme in concentrate/total enzyme in original sample) x 100.

(b) Filtration conducted at 11°C; all other studies were conducted at 25-26°C.

(c) Not a true recovery efficiency, since 47 percent of the β-galactosidase activity of the control sample was lost on standing.

When β-galactosidase was concentrated using the UM-1 membrane, it was possible to achieve a 30.3 volume-concentration factor in approximately 10.5 hours at 60 psig filtration pressure. No enzyme activity was detected in the filtrate. This fact is again within reason, since the molecular weight of the enzyme is estimated to be approximately 850,000. In viewing the recovery efficiency for this enzyme, it appears at first glance to be exceedingly low (24 percent). However, the control sample on standing in the identical buffer system, temperature, and time also showed a drastic loss (47 percent) of enzyme activity. The inherent unstable nature of this enzyme and possible enzyme deposition onto the membrane presumably were responsible for this low recovery efficiency. One therefore sees that, in processing biologically active materials, extreme precautionary measures must be considered in order to maximize recovery efficiency.

The degree of concentration and the recovery efficiencies of trypsin at 11 and 26°C using the UM-1 membrane are also shown in Table 2. At 11°C, a 4.52 volume-concentration factor was achieved in approximately 7 hours, whereas a 4.56 concentration factor was achieved in 6 hours at 26°C. No enzyme activity was detected in the filtrate, which again appears to be reasonable, since its molecular weight is about 24,000. The enzyme recovery efficiencies at 11 and 26°C were, respectively, 91.3 and 76.6 percent. The losses in enzyme activity for trypsin will now be examined and presented in a more detailed manner.

To determine the possible reasons for the enzyme losses during membrane concentration, a series of detailed experiments were performed using the same crude trypsin preparation. The enzyme solutions at 0.15 mg/ml were prepared in tris buffer and ultrafiltered at two temperatures. Approximately 400 ml of the original solutions were concentrated nine-fold and the various fractions were analyzed for their enzyme activity and protein content. The results of these studies are summarized in Table 3. It can be seen that at 26°C essentially all of the protein can be accounted for, but only 81 percent of the enzyme activity was recovered. When one further examines the specific enzyme activities of the original sample, the concentrate, and washings from the membrane, two interesting facts are illustrated: The concentrate is higher in specific activity, indicating that a partial purification was achieved. However, what is discouraging is the substantial reduction in the specific enzyme activity of the washings

TABLE 3. DETAILED ANALYSES OF TRYPSIN ACTIVITY DURING UF PROCESSING WITH A UM-1 MEMBRANE

Temperature, °C	Enzyme Activity Recovered, percent	Protein Recovered, percent	Specific Enzyme Activity, EU/mgP		
			Initial	Concentrate	Membrane
11.5	87	91	45	57	10.2
26	81	105	45	57	2.8

obtained from the membrane. These findings would indicate that, when the enzyme is deposited on the surface of a membrane, an enzyme inactivation process had occurred due to reasons unknown. The same conclusion can be drawn when the results from filtration studies at 11.5°C are examined. However, the inactivation process appears to be less drastic at the lower temperature, as exemplified by the higher recovery efficiency and the higher specific activity of the washings from the membrane. These findings therefore could explain some of the activity losses of the other enzyme systems that were ultrafiltered. The membrane technologists should be aware of this phenomenon and look for possible solutions for its elimination.

The reader will note that ultrafiltration fluxes were not reported in the discussion. Although these data were obtained during experimentation, the authors' intentions were to leave out this information. The reason for this exclusion is due to the laboratory nature of the filtration cell. Presenting the fluxes may be misleading, since no attempt was made to reduce the deposited materials on the membrane. This phenomenon and its elimination presumably will be examined and explained in detail by other authors in this Conference.

Enzyme Purification by Ultrafiltration

From the previous results on the specific enzyme activity (trypsin) after concentration with a reasonable "tight" membrane such as UM-1, it has been shown that some degree of purification can be achieved. In order to explore this phenomenon in more detail, the following analyses and experiments were performed.

A crude enzyme preparation (trypsin) was selected for these studies. This sample was first chromatographed on a G-75 Sephadex gel. The results of the liquid chromatogram are shown in Figure 3. The eluates from the gel column were analyzed for protein and enzyme activities. It can be seen from this diagram that the enzyme activities of the eluates corresponded to very low protein contents (elution volumes 45 to 65 ml). Furthermore, the bulk of the protein impurities appeared in eluate volumes ranging from 72 to 95 ml. These protein fractions contained no enzyme activity and ranged in molecular weight from 11,000 to 19,000 as determined previously through the calibration of the gel. Since the enzyme (trypsin) has a molecular weight of about 24,000, it appeared possible to select a UF membrane which would pass the protein impurities and reject the enzyme. Based on these facts, the HFA-200 membrane (Abcor, Inc., Cambridge, Mass.) was selected, since the nominal molecular-weight cutoff was about 20,000.

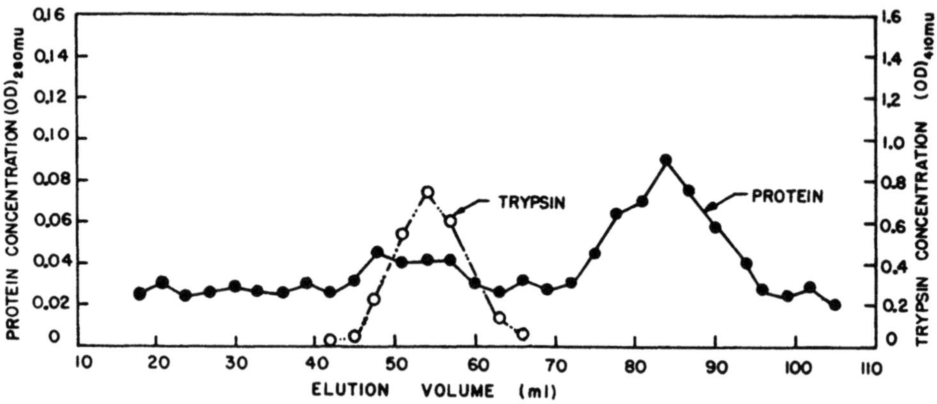

Figure 3. Gel filtration of trypsin solution before ultrafiltration (0.15 mg/ml trypsin in 0.1M, pH 8.2, tris buffer solution).

The quantitative results of these ultrafiltration experiments are summarized in Table 4. First, 1420 ml of the crude trypsin was concentrated to a volume of 170 ml, representing a concentration factor of 8.35. Due probably to the diffused nature of the molecular-weight cutoff for the HFA-200 membrane, a portion of the trypsin and the low-molecular-weight impurities were able to pass through the membrane. The enzyme assays showed that 70.2 and 20.0 percent of the initial activity were found in the concentrate and filtrate, respectively. The 9.8 percent enzyme unaccounted for presumably was inactivated similar to that previously reported for the UM-1 membrane. Accompanying the concentration, it can be seen from Table 4 that, second, a purification factor of 2.35 also was achieved. This is quite conceivable, in view of the elution diagram, which had demonstrated low-molecular-weight protein impurities. This purification was further substantiated and is more dramatically shown in the gel-elution diagram of the concentrate in Figure 4. The results show that a significantly higher concentration ratio of the trypsin protein to the impurity protein exists in the concentrate, indicating the removal of lower-molecular-weight impurities. It is of interest to note that the impurities which passed through the membrane still possessed a range of molecular weights from 11,000 to 19,000. However, if one superimposes the elution diagrams of Figures 3 and 4, it can be seen that the concentrate contains a much larger fraction of the higher-molecular-weight impurities. These findings would reinforce the general nature of a diffused molecular-weight cutoff, which can be anticipated with ultrafiltration membranes. Last, a gel chromatogram of the filtrate also was prepared and is shown in Figure 5. As one would expect, the major components which passed through the membrane consisted of low-molecular-weight compounds.

From the results presented in these studies, it can be seen that concentration and partial purification of a crude enzyme preparation can be readily achieved in a single-pass UF process. It can also be concluded that, with quantitative knowledge of the impurities in a sample along with general membrane-rejection characteristics, it is possible to select a proper membrane for a given task. There are probably many industrial and research enzymes for whose overall recovery processes UF membranes could conceivably play an important role. With the introduction of commercial-scale ultrafiltration units, this process undoubtedly will become more important in biological processes in general.

TABLE 4. CONCENTRATION AND PURIFICATION OF TRYPSIN USING AN HFA-200 CELLULOSE ACETATE MEMBRANE

Initial trypsin concentration = 0.15 mg/ml in 0.1 M, pH 8.2, tris buffer containing 0.02 M $CaCl_2$.

	Before Concentration	Concentrate	Filtrate
Volume, ml	1,420	170	1,250
Concentration factor	--	8.35	--
Total enzyme, enzyme unit	10,900	7,650	2,180
Enzyme recovered, percent	--	70.2	20.0
Specific enzyme activity, EU/mgP	45	106	14.0
Purification factor of concentrate	--	2.35	--

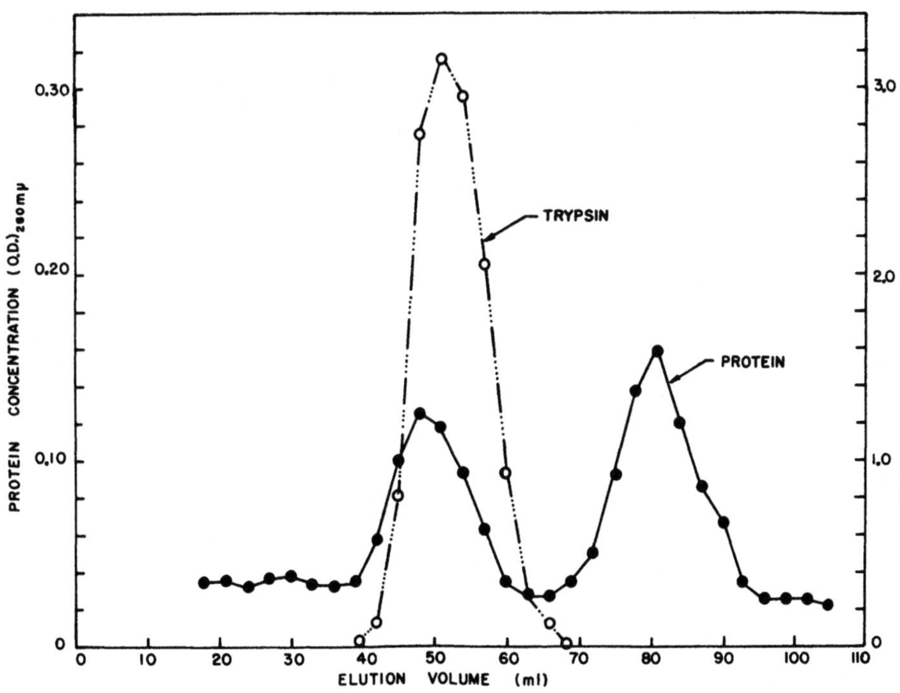

Figure 4. Gel filtration of trypsin solution after concentration with an HFA-200 membrane.

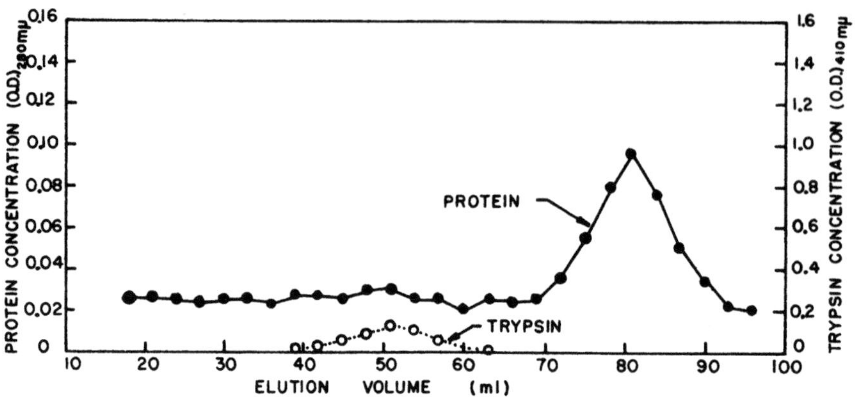

Figure 5. Gel filtration of trypsin filtrate after passing through an HFA-200 membrane.

Membrane Fermentor

In batch fermentations, the kinetics of growth for microorganisms generally can be differentiated into three distinct phases: a lag phase, a logarithmic growth phase, and a stationary phase. When one is considering the production of primary or secondary metabolite by an organism, the absolute value of the cell concentration ultimately reached in the stationary phase often dictates the absolute concentration of the metabolite excreted. One is, therefore, constantly searching for means of maximizing product concentration.

Qualitatively, the reasons why microorganisms approach a stationary phase during their growth cycle are often given as the depletion of one or more essential nutrients and/or formation of toxic metabolic production, either of which could curtail further growth. However, the quantitative identification of those compounds responsible for growth curtailment is often a difficult if not impossible task. It therefore was hypothesized that the coupling of an ultrafiltration membrane with a fermentor during cultivation of an organism may partially eliminate the role of toxic metabolite formation by the organism. One can envision the function of the "membrane fermentor" in the following manner.

Let us assume that, when an organism reaches its stationary phase due to buildup of toxic products, we continuously feed fresh substrate to this culture. Simultaneously, when a UF membrane is coupled to the fermentor, some of the toxic products now can be continuously removed. The membrane in effect acts similarly but not completely as a centrifuge for recycling of the microbial cells. One main difference between the UF membrane and a centrifuge is that macromolecules produced extracellularly in the broth can also be retained by the membrane. Therefore, if one is interested in the production of an enzyme, the membrane fermentor will simultaneously concentrate and perhaps purify the desired end product and thus facilitate later recovery operations. Under these assumptions, a membrane fermentation was carried out. The procedure and apparatus for this system have already been described.

The results from the membrane fermentor for the cultivation of *Clostridium histolyticum* are shown graphically in Figures 6 and 7. Substrate feed was added to the fermentor at approximately the twelfth hour of fermentation. At this time, the organism was about to enter its stationary phase. From Figure 6, it can be seen that the cell weight in the control fermentation reached a plateau at about 4.0 gm of dry cells per liter of culture broth. However, in the membrane system, the cells continue to divide and reach a final cell concentration slightly greater than 10 gm of dry cells per liter.

The enzyme production in the membrane fermentor and the control experiment are shown graphically in Figure 7. In a manner similar to cell growth, enzyme activity in the control system attained a maximum value of 180 enzyme units per ml of culture broth. In the control system, the enzyme activity began to decrease with time, presumably due to an unfavorable environment. On examining the enzyme activity of the membrane fermentor, it can be seen that the protease continued to be excreted up to concentration of 700 enzyme units per ml of broth. The filtrates from the fermentor were analyzed for enzyme activities and none were found during the course of the fermentation. This finding is anticipated, since the HFA-300 membrane is reported to reject molecular weights up to 50,000, while the enzyme's molecular weight is nearly twice this value.

An interesting comparison can be made from these studies with respect to the ratio of the enzyme excreted per unit mass of cell produced. In the control experiment, the maximum ratio was 45 enzyme units per mg of cell dry weight. However, for the membrane system this ratio increased to 70 enzyme units per mg of cell dry weight. This would indicate that, under a more favorable environment in the membrane fermentor, the organism is more "efficient" in its biosynthetic machinery in the excretion of the enzyme. The results presented here thus far do not offer concrete proof as to the nature and type of toxic metabolites which may have been present or removed through ultrafiltration. However, the end results of the membrane fermentor are the increases in cell and enzyme concentrations which were achieved. These increases are significant since, by using a membrane with a fermentation, the productivity of a given fermentor can be substantially increased. Furthermore, a possible reduction in efforts of the subsequent recovery operations may result. Conceivably, there are other types of fermentation in which benefits may be derived using a similar approach.

Membrane Reactor

General Discussion. The use of enzymes, besides as an additive in laundry detergents, is quite extensive at the laboratory as well as at the industrial scale. For example, the saccharification of starch in grain by α-amylase and glucoamylase is commonly employed in grain spirits production. Both of these enzymes are also used in the commercial production of glucose as well as syrups. In addition, there are many organic transformations which employ the specific action of enzymes for their conversion. In these categories are the steroids and the synthetic penicillins. Undoubtedly, many other useful enzyme conversion processes will be developed in the days to come in view of man's ability to synthesize the enzyme molecule.

Enzymes are biological catalysts. Therefore, during a reaction, the enzyme need only be present in catalytic amounts. Mathematically, a simplified enzyme kinetic model is often expressed in the following manner:

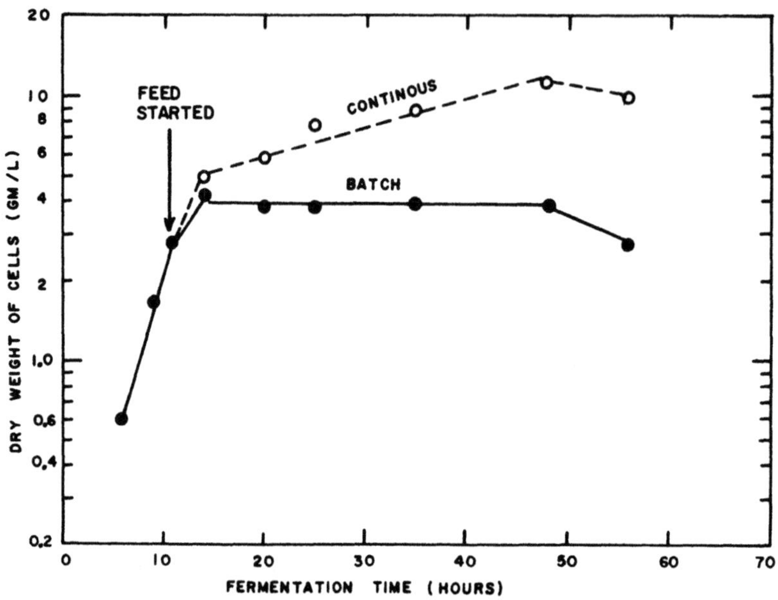

Figure 6. Comparison of batch and continuous membrane fermentor on cell production (T = 37 C, HFA-300 membrane, ΔP = 100 psi).

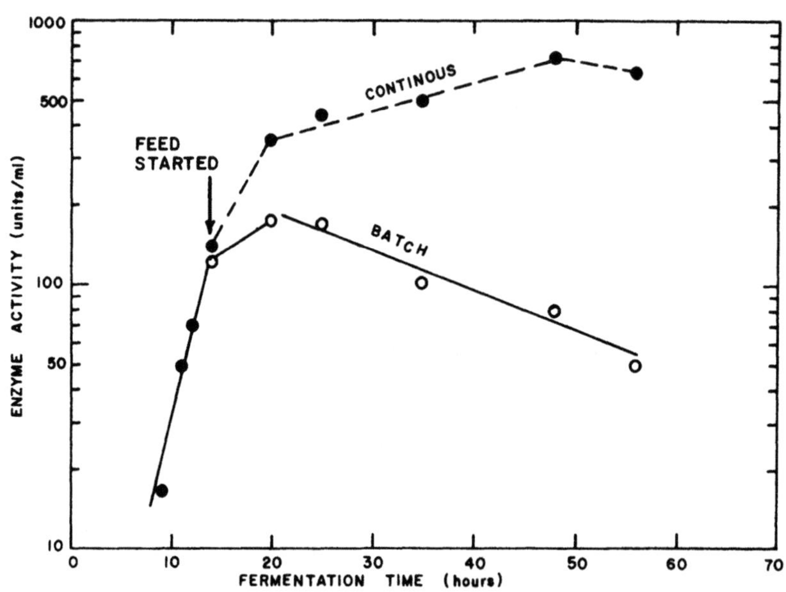

Figure 7. Comparison of batch and continuous membrane fermentor on enzyme production (T = 37 C, HFA-300 membrane, ΔP = 100 psi).

$$E + S \rightleftharpoons ES$$

$$ES \rightarrow E + P$$

where

E = enzyme concentration

S = substrate concentration

ES = enzyme-substrate complex concentration

P = product concentration

In theory, therefore, it should be possible to convert a very large amount of substrate with a small but finite amount of enzyme if reutilization of the catalyst can be accomplished. In chemical catalysis, such as in "cat-cracker" operation, the catalysts are indeed regenerated for further reuse. It is the intent of this section of the paper to examine the technical feasibility of enzyme reutilization in continuous enzymic reactions.

In theoretically analyzing the reutilization of a biological catalyst, there are many possible approaches. These include insolubilizing an enzyme through adsorption or chemically bonding it onto an inert matrix or the use of conventional enzyme-recovery systems such as liquid chromatography, liquid partitioning, and the like. However, when one examines the nature of the reactants, products, and catalysts, one obvious and simple method would be the use of an ultrafiltration membrane. This was the method which we selected; the manner of operation will be briefly described.

A reactor containing an enzyme preparation is continuously fed with substrate as illustrated schematically in Figure 2. This reactor also contains a UF membrane that will allow the passage of products of reaction while retaining the enzyme. The findings on two enzymes' systems now will be presented and discussed.

Enzyme Retention in Membrane Reactor. When using ultrafiltration membranes in conjunction with enzymic reactions, the membrane and enzyme must be compatible so that a true steady state can be achieved. The criteria for this condition must be the degree of enzyme retention and the rate of enzymic reaction. The membrane-retention characteristics will now be presented for the enzyme systems that were studied.

Shown in Figure 8 are the enzyme-retention characteristics for α-amylase using the membrane reactor. The concentration of the enzyme was always 0.16 mg/ml, with the substrate to the reactor a 0.4 percent starch solution. The volume of filtrate passed through the membrane was normalized as V_f/V_o in order to have a general means for comparison. In this case, the symbols are defined as

V_f = volume of filtrate passed

V_o = volume of liquid inside the cell (constant) .

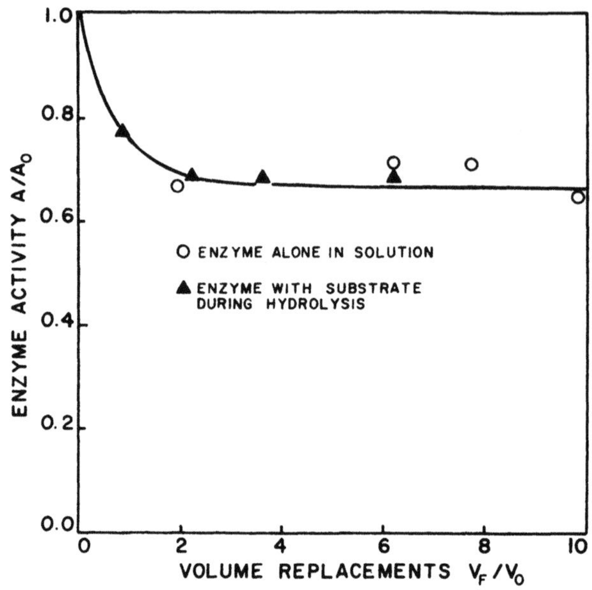

Figure 8. Retention of α-amylase in the continuous membrane reactor (0.16 mg/ml α-amylase, PM-10 membrane, ΔP = 15 psi, T = 40 C).

Although the molecular weight of α-amylase (50,000) is much greater than the nominal-molecular-weight cutoff of the membrane (PM-10 = 10,000), it can be seen that, during the initial portion of the ultrafiltration process, some of the enzyme passed through the membrane. The retention characteristics for the enzyme in Figure 8 are expressed as the ratio of the enzyme activity in the cell during filtration to the initial enzyme activity. During the transient period, for $V_f/V_0 < 2.0$, approximately 35 percent of the initial activity was lost due to the low membrane-rejection efficiency. This enzyme loss was experienced regardless of the presence or absence of the substrate. However, after two volumes had been replaced, the rejection efficiency increased to 100 percent and no further enzyme loss was encountered.

The same retention phenomenon was also noted for glucoamylase at an initial concentration of 1.83 mg/ml; this is illustrated graphically in Figure 9. This enzyme is similar in molecular weight (ca. 50,000) to that of α-amylase. It should be pointed out that the decrease in enzyme activity in the filtration cell was primarily due to physical loss by passage through the membrane. This conclusion is reached since the analyses of the enzyme activity of the filtrates showed that only about 5 percent of the enzyme had been inactivated (see data of Figure 9). In either case, it was fortunate that, at "steady state", 100 percent enzyme rejection occurred, which renders the membrane reactor practical to operate.

The passage of the enzyme during the transient period was not explored analytically in detail. One can speculate, however, as to what may have occurred. It could be conceived that, during the initial period, the enzymes and/or protein impurities in solution were convectively transported to the membrane surface. The presence of this slow-moving hydrogel on the membrane surface could act as a dynamically formed secondary membrane. This "membrane" could then act as an additional barrier to enzyme transport and thus control the retention

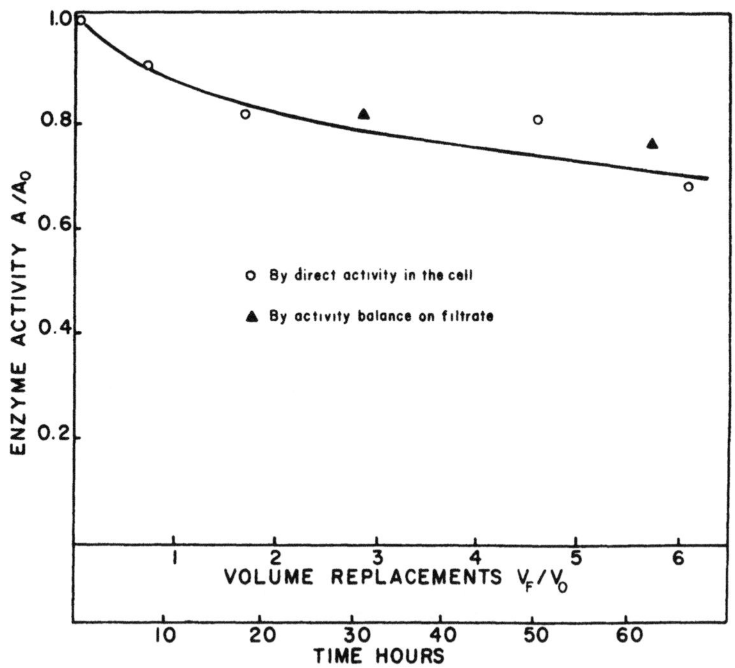

Figure 9. Retention of glucoamylase in the continuous membrane reactor (1.83 mg/ml glucoamylase, PM-10 membrane, ΔP = 15 psi, T = 40 C).

characteristics as shown in the results. Alternatively, there may be small molecular-weight fragments in the enzyme preparation which are "biologically active". These compounds are able to pass through the membrane and thus account for the initially low membrane-rejection efficiencies. Further studies, however, must be performed in order to substantiate or refute either of these hypotheses.

Continuous Enzymic Reaction With the Membrane Reactor. The results from a typical membrane-enzyme reactor using α-amylase for the hydrolysis of starch are shown in Figure 10. The reaction was conducted continuously for approximately 4 days, producing over 3 liters of product. Other studies have been performed over periods as long as 12 days, with no loss in enzyme activity after "steady-state" rejection had been achieved. The temperature of reaction was 40°C and the pressure for filtration maintained at 15 psig. The average dextrose equivalent (DE) of the product was 34 during the steady-state periods (DE = 50 for maltose).

It can be seen from Figure 10 that the carbohydrate concentration in the retentate increased almost linearly with time from 0.4 to about 0.9 percent. This increase is due to two phenomena occurring simultaneously: partial rejection of the reaction products and passing of enzyme through the membrane. The partial rejection of the hydrolysis products can best be illustrated if we examine the Sephadex gel chromatogram of the various fractions in Figure 11. It can be seen that the hydrolysis products (retentate) contain dextrins with a broad distribution of molecular weights ranging downward from 20,000. Since the UF membrane passes

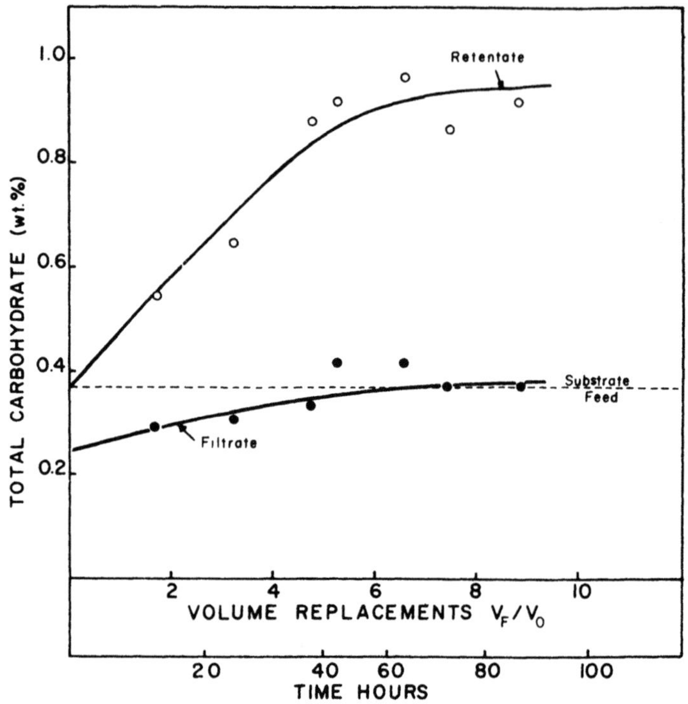

Figure 10. Continuous hydrolysis of starch by α-amylase in a membrane fermentor (feed conc. = 0.4 percent, T = 40 C, PM-10 membrane, ΔP = 15 psi, enzyme = 0.16 mg/ml).

molecular-weight materials below 10,000, an accumulation of the large dextrins is inevitable. Thus, an unsteady-state situation existed in the reactor until approximately five volume replacements had occurred. After this time, a true steady state resulted, as exemplified by the constant levels in concentration for the feed, retentate, and filtrate (and previously shown enzyme level).

The loss of enzyme through the membrane also contributed to the carbohydrate concentration increase during the nonsteady-state portion of the hydrolysis. This decrease in the enzyme activity resulted in a lower overall reaction rate and thus caused the substrate to accumulate in the cell. This phenomenon, however, played an active role only up to two volume replacements.

The second enzyme system that was used in the membrane reactor was the action of glucoamylase on starch. The reaction products from this enzyme, essentially pure dextrose, are different from α-amylase hydrolysis. The initial glucoamylase concentration was 1.83 mg/ml, an excessive amount of enzyme for completely hydrolyzing a starch feed concentration of 8 percent. The results from these experiments are shown in Figure 12. Due to the excess enzyme in the reactor and the formation of low-molecular-weight (glucose) hydrolysis product, steady state was attained much more rapidly when compared with the α-amylase reaction. Furthermore, the difference in the carbohydrate concentration between the retentate and

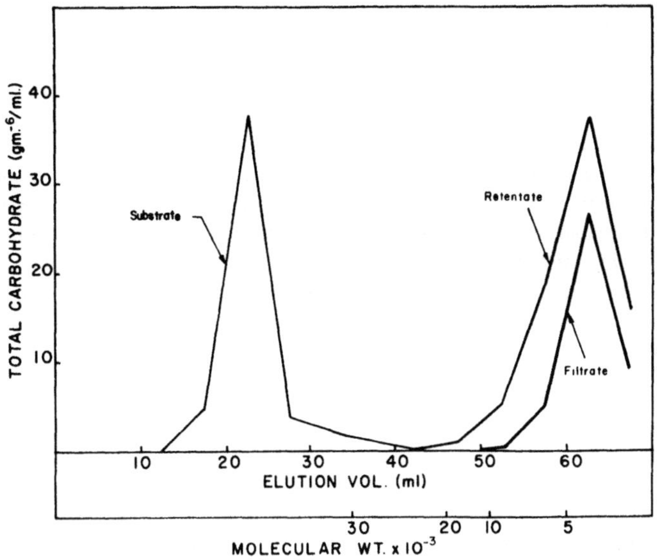

Figure 11. Sephadex gel chromatogram of various fractions from the membrane reactor.

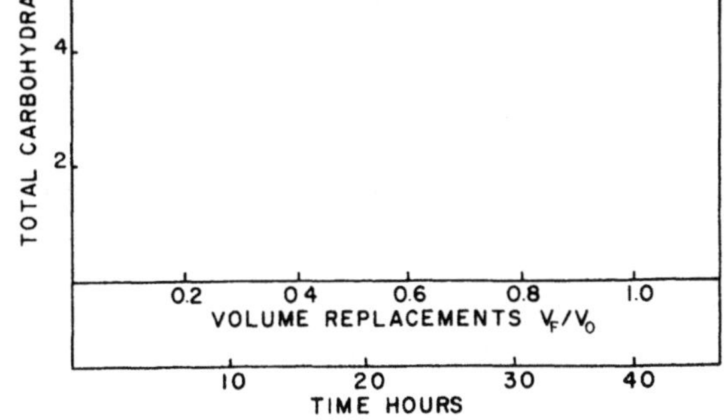

Figure 12. Continuous hydrolysis of starch by glucoamylase in membrane reactor. (Feed conc. = 8 percent, T = 40 C, PM-10 membrane, ΔP = 15 psi, init. enzyme conc. = 1.83 mg/ml).

filtrate was only about 1 percent. The results again demonstrate the feasibility of the membrane reactor in carrying out continuous enzymic reactions and reutilization of the enzyme.

Although both enzyme systems demonstrated the ability of a membrane reactor for the reutilization of enzymes, there is one subtle difference between the two. This is the ability of the membrane during α-amylase hydrolysis to "fractionate" the products of reaction. One can conceive the idea of a membrane reactor in which, by controlling the enzyme-substrate ratios, reaction time, and membrane selection (in terms of molecular-weight cutoff), it may be possible to control the molecular-weight distribution of the reaction products. On the other hand, for enzymic reactions in which the substrate and products are low in molecular weights, a properly designed membrane reactor still can be made operative through correct manipulation of the enzyme level and reaction time. It is the authors' belief that the findings presented in these studies should open many doors in other enzyme conversion processes.

REFERENCES

Anon (1969A), "Frenzy Over Enzymes", *Chemical Week,* **104** (11), 75.

Anon (1969B), "Enzymes Invade Laundry Products Market", *Chem. & Eng. News,* **47** (5), 16.

Ashwell, G. (1957), (12) "Colorimetric Analysis of Sugars", in *Methods in Enzymology,* Colowick, S. P., and Kaplin, N. O., eds, Vol III, p 85, Academic Press, N. Y.

Bernfeld, P. (1951), "Enzyme of Starch Degradation and Synthesis", in *Advances in Enzymology,* **12**, 379.

Berman, S., Lowenthal, J. P., Webster, M. E., Altieri, P. L., and Gochenour, R. B. (1961), "Factors Affecting the Elaboration by *Clostridium histolyticum* of Proteinases Capable of Debriding Third Degree Burn Eschars on Guinea Pigs", *J. Bacteriol.,* **82**, 582.

Blatt, W. F., Feinberg, M. P., Hopfenberg, H. P., and Saravis, C. A. (1965), "Protein Solutions: Concentration by a Rapid Method", *Science,* **150**, 224.

Blatt, W. F., Hudson, B. G., Robinson, S. M., and Zipilivan, E. M. (1967A), "Fractionation of Protein Solutions by Membrane Partition Chromatography", *Nature,* **216**, 511.

Blatt, W. F., Robinson, S., Robbins, A., and Saravis, C. A. (1967B), "An Ultrafiltration Membrane for Resolution and Purification of Bovine α-lactalbumin", *Anal. Biochem.,* **18**, 81.

Charney, J., and Tomarelli, R. M. (1947), "A Colorimetric Method for the Determination of the Proteolytic Activity of Duodenal Juice", *J. Biol. Chem.,* **171**, 501.

Citri, N. (1964), "Determination of Penicillin Activity", *Methods Med. Res.,* **10**, 221.

Garnot, P. O. (1966), "On the Heterogeneity and Purification of Commercial Trypsin Preparations", *Acta Chem. Scand.,* **21**, 175.

Kay, C. M., Smillie, L. B., and Hilderman, F. A. (1961), "The Molecular Weight of Trypsinogen", *J. Biol. Chem.*, **236**, 118.

Hu, A.S.L., Wolfe, R. G., and Reithel, R. J. (1959), "The Preparation and Purification ofβ-Galactosidase From *Escherichia coli*, ML 308", *Arch. Biochem. Biophysics,* **81**, 500.

Keller, S., and Mandl, I. (1963), "The Preparation of Purified Collagenase", *Arch. Biochem. Biophysics,* **101**, 81.

Michaels, A. S. (1968A), "Ultrafiltration", in *Advances in Separations and Purifications,* E. S. Perry, ed, John Wiley, New York, N. Y.

Michaels, A. S. (1968B), "New Separation Technique for the CPI", *Chem. Eng. Prog.*, **64** (12), 31.

Pattison, D. A. (1968), "Membranes Complete For Separation Markets", *Chem. Eng.*, **75** (12), 38.

Penhoet, E., Kochman, M., Valentine, R., and Rutter, W. J. (1967), "The Subunit Structures of Mammalian Fructose Diphosphate Aldolase", *Biochem.*, **6** (9), 2940.

Rickles, R. N. (1967), "Membrane Technology and Economics", Noyes Development Corp., Park Ridge, N. J.

Seifter, S., Dayton, S., Novic, B., and Muntwyler, E. (1950), "The Estimation of Glycogen With the Anthrone Reagent", *Arch. Biochem. Biophysics,* **25**, 191.

Schlomo, H., Feingold, D. S., and Schramn, M. (1955), in "Methods of Enzymology", Colowick, S. P., and Kaplin, N. O., Eds, Vol I, 241, Academic Press, New York.

Wang, D.I.C., Sonoyama, T., and Mateles, R. I. (1968), "Enzyme and Bacteriiphase Concentration by Membrane Filtration", *Anal. Biochem.*, **18**, 277.

Zipilivan, E. M., Hudson, B. G., and Blatt, W. F. (1969), "Separation of Immunological Active Fragments by Membrane Partition Chromatography", *Anal. Biochem.*, **29**.

The authors wish to acknowledge N.I.H. Training Grant No. TO1-ES00063-04 and Grant No. UI-00723-02 from the Department of Health, Education and Welfare, Public Health Service, whose partial support made this work possible.

A CONSIDERATION OF THE PARAMETERS GOVERNING MEMBRANE FILTRATION – PARTICULARLY OF PROTEINACEOUS SOLUTIONS

C. T. Badenhop, A. T. Spann, and T. H. Meltzer
Gelman Instrument Company
Ann Arbor, Michigan

INTRODUCTION

That the pore characteristics of membranes govern their filtrative behavior seems rather axiomatic. Yet, the view that filtration is solely a sieve-type phenomenon is known to be simplistic. As the pore sizes diminish, the nature of the polymer constituting the membrane matrix contributes increasingly to the possibility of interaction between the filter and what is being filtered. Additionally, such structural features as the pore-size distribution, the shapes of the pores, the presence of pore-size gradients across the membrane thickness, and anisotropic effects in general all influence the filtration properties of membranes.

In this paper, consideration is given to these several factors in a study based on aqueous proteinaceous solutions. Some assessment of their individual influences is made, and the prediction of membrane performance in beer filtration is ventured on the basis of computerized regression analysis.

NEW APPARATUS FOR PORE-SIZED DISTRIBUTION

Although membrane filtration is a complex phenomenon, the single most important influence governing its performance is its porosity – the number and size of the filter pores. Several methods exist for determining pore size, but most are too unwieldy or too time consuming to adapt to ready usage. There is in existence an ASTM method, D-2499, that describes a procedure for graphically determining the flow pore distribution of a membrane. However, this technique introduces large errors in the generation of pore histograms owing to the difficulties of transposing graphical data. Furthermore, even these representations are time consuming to prepare and do not yield a continuous distribution function. The need for a better method has led us to the development of an instrument that is capable of performing the complete analysis of flow pore distribution of filter media in the 10.0-0.25μ-size range with essentially none of the operator error inherent in D-2499. This equipment, coupled with laboratory tests of bacterial retention and other particulate studies, can characterize the performance of membrane filters with significantly improved reliability.

Description of the Test Method

Method D-2499 discusses the essential relationship between pressure, surface tension of a wetting fluid, and pore diameter. This relationship predicts that a wetted membrane with applied differential air pressure will permit no flow at pressures below a critical level referred to as the "bubble point". At the bubble point, wetting fluid is forced out of the smallest pores and flow thus begins. Increasing the air pressure will cause the next larger pores to open, and so on, until the air pressure is sufficient to open all the pores of the membrane. These pores are opened in such a fashion as to form an inversely proportional relationship between pore size and the pressure required for evacuation. Appropriate mathematical analysis of this phenomenon yields the pore distribution of the membrane.

Since it is difficult to extract the number of actual pores from these data, variables are grouped together to yield pore-size measurements that may be used to assign percentages of flow accounted for by two or more preselected ranges of pore sizes. One of the statistics thus generated is the mean flow pore (MFP), which is defined as the pore size for a given membrane whose magnitude is such that 50 percent of total flow is due to pores smaller than the MFP and 50 percent of flow is due to pores larger than the MFP. The percentage breakdown can be extended to any number of discrete intervals so that a bar-graph histogram of flow pore distribution is derived. The equation (see Figure 1) for percent of flow for a discrete pore-size interval, I, is:

$$\frac{W_2}{D_2} - \frac{W_1}{D_1} \times 100 = \%$$

where W_2 and D_2 are wet and dry flow rates, respectively, at the high-pressure (low-pore-size) side of the interval and W_1 and D_1 are wet and dry flow rates, respectively, at the low-pressure (high-pore-size) side of the interval. With pore size measured in microns (10^{-6} meters), pressure, P, measured in pounds/square inch, and with kerosene as the wetting fluid, the relation between the pore size and pressure (from ASTM D-2499) is: $\mu = \frac{12.5}{P}$. This equation is not a precise relation for all materials, as the contact angle between the kerosene and a given type of membrane may not always be the same. However, within any polymer group the only change in the relation is the value of the constant. This has been checked for all the membranes used at Gelman Instrument Company and found to be reasonably accurate. To find the number of pores that exist in any range of pore sizes, a relation between the pore size and some other measured reference is necessary. The wet flow curve offers such a reference.

If one considers the deviation between the wet flow curve and the dry flow curve (see Figure 1), it is seen that the wet flow is always less than the dry flow but approaches the dry flow asymptotically at high pressures. The difference between the curves is due to the interference with flow caused by the surface tension of the fluid blocking all holes that are smaller than those which are represented by the relation $\frac{12.5}{P}$, where P is pressure under

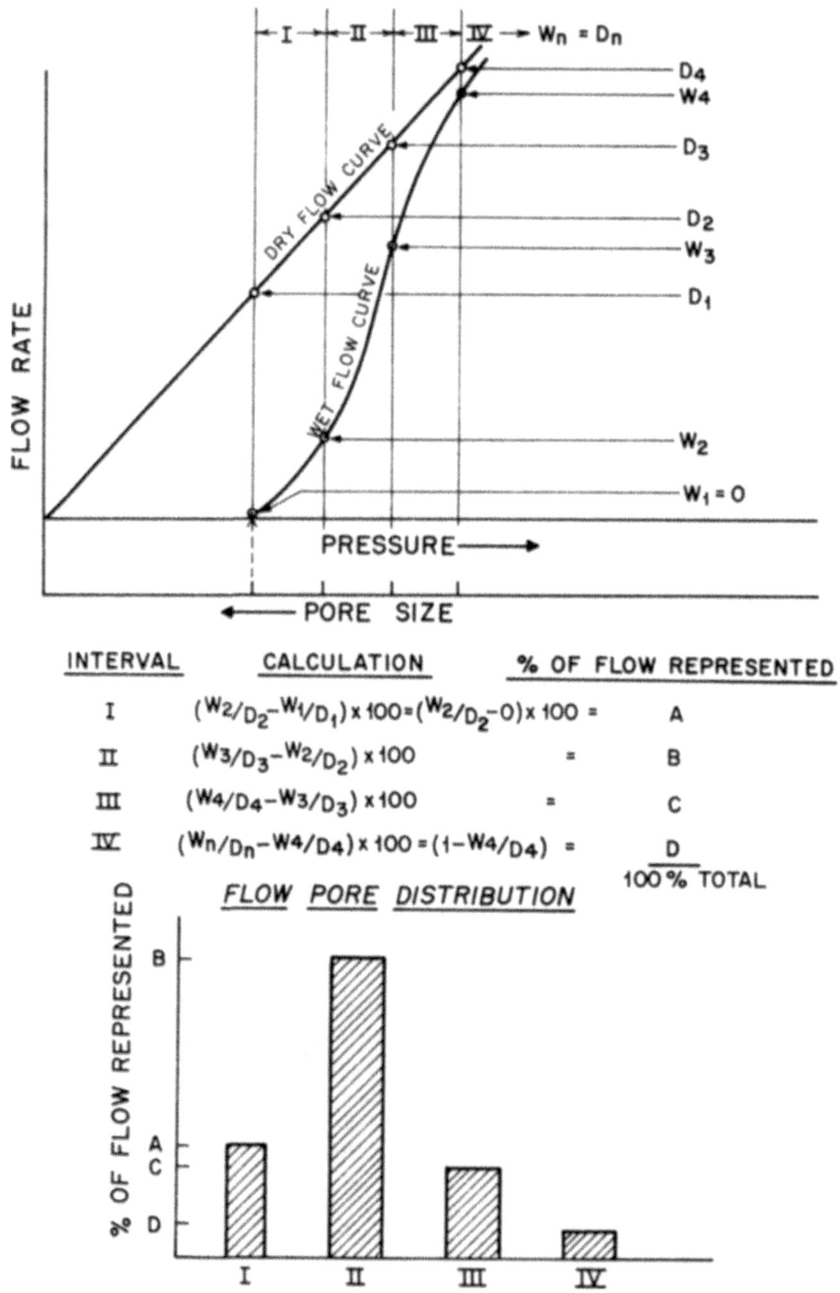

Figure 1. Manual calculation of flow pore distribution.

consideration. The total wet flow at a point P_i is due only to those holes larger than $\frac{12.5}{P_i}$ microns.

By dividing the wet flow curve by pressure, a function is obtained which extracts the effect of increased flow through the open pores due to increasing pressure. This new curve represents the change of flow with respect to pressure only due to change in number of flow pores opening. Figure 2 is the curve of wet flow/pressure. We are interested in the number of pores which exist in any interval between P_1 and P_2 on the pore axis between BP (bubble point) and some point P_n. The quotient $\frac{Q_2-Q_1}{\epsilon}$ represents the number of flow pores per unit interval in size where $Q = \frac{\text{wet flow}}{P}$. If we let $\epsilon \rightarrow 0$, then $Q_2 \rightarrow Q_1$ and $\frac{Q_2-Q_1}{\epsilon} \rightarrow D_pQ$.

Therefore, the derivative of the function Q with respect to pressure is the desired relation. However, the differentiation of Q with respect to time is the function obtained by dividing D_tQ by D_tP. If D_tP is a constant, no problem exists, since it can be grouped with all constants. If D_tP is not a constant, an error in relative numbers of pores will occur with respect to pore size. The D_tP is generated by the pressure program section of the equipment and was found to be somewhat nonlinear. The degree of error caused by this nonlinearity, however, is very small for actual filters, because the total range of porosity is small enough so that the D_tP over the pressure range required to display the pore distribution may be considered constant.

Description of Apparatus

The basic problems of analyzing flow pore distribution of filter media include measuring the air flow through a wetted membrane as a function of pressure and relating this flow to the

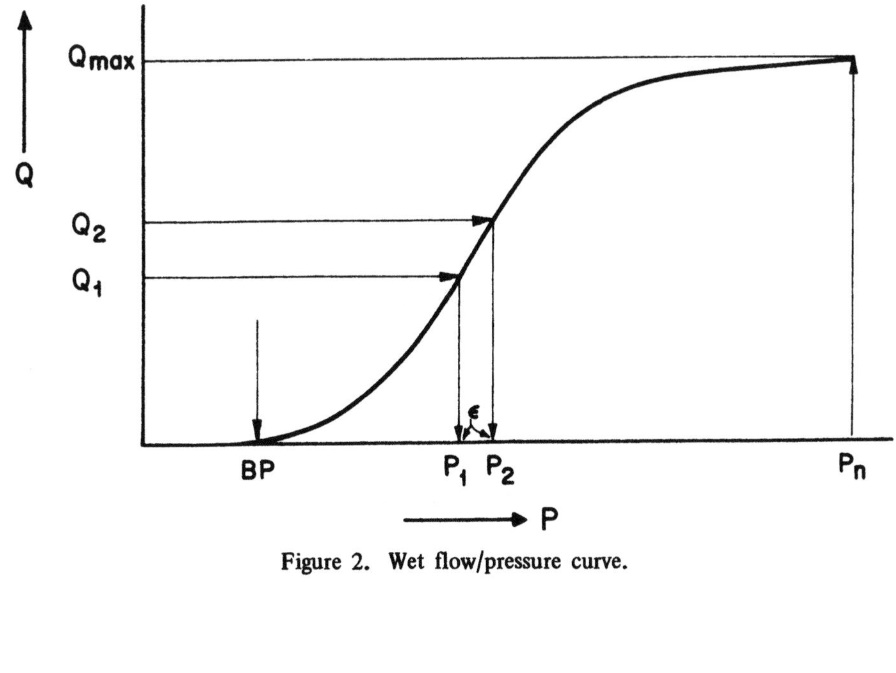

Figure 2. Wet flow/pressure curve.

air flowing through an equivalent dry membrane as a function of pressure, so that the ratio of wet/dry flows may be obtained. The wet/dry flow function then is differentiated with respect to time to generate the flow pore distribution curve. Since the ratio curve can only be differentiated with respect to time, it is necessary that the rate of increase of pressure be linear with respect to time or very reproducible, with a known deviation from linearity. The required system for establishing the desired flow pore distribution curve is

(1) Programmed linear pressure increase

(2) Air-flow measurement systems that yield an analog readout

(3) Analog systems to divide wet air flow by dry air flow and to differentiate the resulting function

(4) Recorders to plot the necessary responses generated.

Pressure Program System. To achieve a linear pressure across the filter membrane, a pneumatic network is used incorporating a pilot relay, variable orifice, and capacity tank (see Figure 3).

The pilot relay is biased to +1 psi, causing a pressure drop of one psi to be applied across the variable orifice. Since no flow other than that necessary to fill the capacity tank occurs, a relatively linear pressure increase is obtained, limited only by the supply pressure. During this

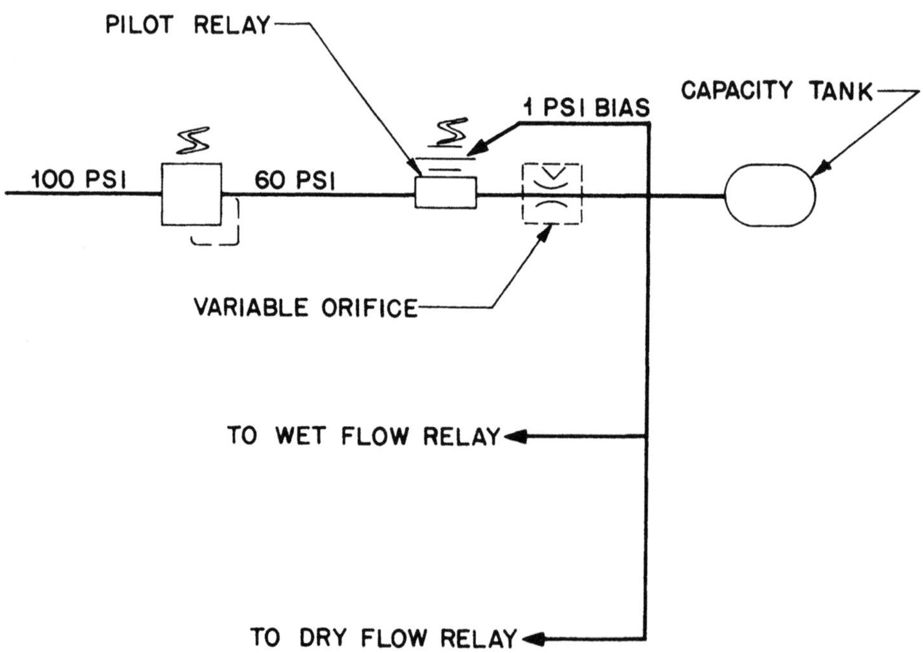

Figure 3. Pressure programmer.

time, a rising pressure signal is being applied to the other pilot relays. Here, there is a slight error in the pressure program due to the compression (increase in density) of the air flowing through the orifice and the resulting change in Reynolds numbers across the orifice.

Flow-Measurement System. Since the range of flows encountered in the study of membranes is very broad, the choice of a measurement device that will cover this range and also provide an analog response is critical. To fulfill these requirements, the Foxboro internal-orifice differential-pressure transmitter was chosen. This instrument can measure a wide range of flows through the use of interchangeable internal-orifice plates. The top works of the system transmits a pneumatic signal proportional to the pressure drop across the orifice plates. By choosing the correct orifice, the desired range of flows can be transmitted. Flow measurements of a compressible fluid must either take into account changes in the pressure of the flowing stream or must be made at constant pressure. The system used in this device (see Figure 4) measures the flow at constant pressure.

From Figure 4, it is seen that a 60-psi main supply is connected through the orifice meter to the high-pressure side of a pilot relay. These lines are sufficiently large and the supply volume available is great enough to maintain 60-psi pressure to the relay at all flow rates that are encountered. The pilot relay receiving the programmed pressure signal reduces the 60-psi line to this signal value, which is applied at high-flow volume across the filter membrane under test. Two identical systems are used – one for the wet filter and one for the dry filter. Since they both receive the same pressure program, identical pressure drops across the wet and dry filter occur.

Analog System. The flow-rate data from the flow meters of both wet and dry flows are in the form of a pneumatic analog signal. Leaving the differential-pressure transmitters, they first must be made linear, since the pressure drop through an orifice is proportional to the square of the flow. These data are rectified by two square-root extractors. The resulting linear signals then are fed to a division computer where the wet/dry ratio is obtained (see Figure 5). The ratio signal then goes to a pneumatic computer, where the derivative of the ratio with respect to time is developed.

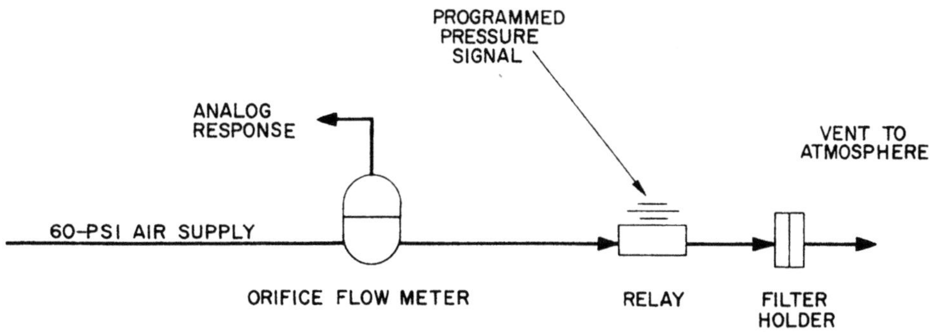

Figure 4. Flow meter schematic.

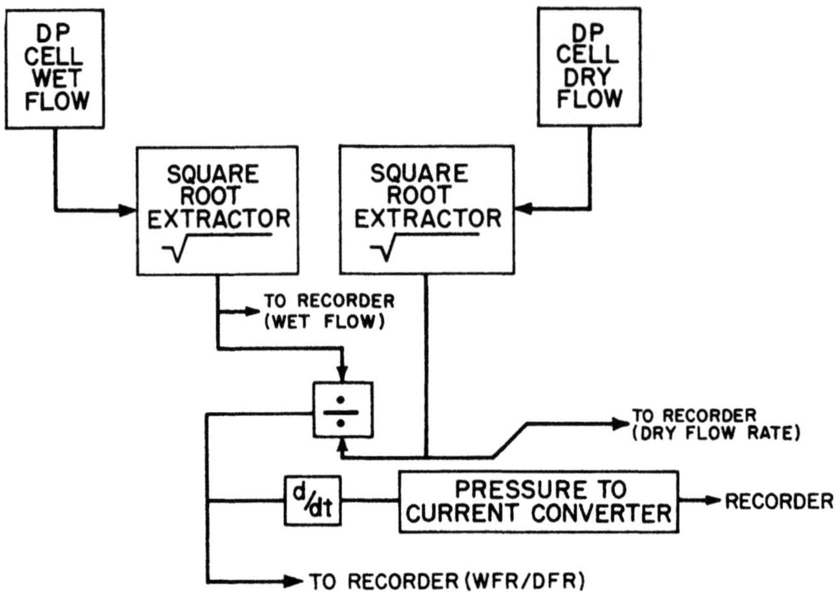

Figure 5. Analog signal path.

Recorders. All the information generated by the analyzer is recorded by two instruments. One is a three-pen polar-coordinate recorder that records the pressure across the filter, the dry air flow, and the ratio of wet/dry flow. The second instrument records the derivatives of the wet flow/dry flow ratio curve.

Calibration of Instrumentation

The entire system must be calibrated to ensure that the responses obtained are meaningful. Calibration of the pressure gauges and flow meters can be done using instruments of known accuracy. The division network can be checked for accuracy by applying known flow rates at both flow meters and calculating the ratios. This instrument shows highly repeatable performance with respect to pressure program and flow reproducibility.

The flow meters must be matched exactly so that identical responses are obtained from both systems. The systems are referenced by use of two calibrated rotameters attached to the filter ports. The pressure signals at the exit ports are measured with 0.25-percent-accuracy pressure gauges to within ±0.05 psi. The flows through the rotameter are checked against the dry- and wet-flow-pen readings. The spans of the flow meters are set at maximum flow readings. The flows then are reduced to zero to readjust the zero point. Flow then is reset at maximum and this adjustment is reset if necessary. This process is repeated until both instruments read the same. Since the flow range of the orifice in the meter is linear within its calibrated range, any nonlinearity in flow reading across the range must be due to recorder inaccuracy. Pen-linkage adjustments are made to correct such aberrations.

The analog computers are calibrated by a series of tuning adjustments that are carefully performed and tested to indicate correct response. The division network is checked, with the flow meters held at constant flow to verify their calibration. Both pen-reading and output-pressure response must be correct, since this signal is further differentiated to yield the pore distribution.

The derivative computer is calibrated for gain and frequency response. An impulse signal is applied to the computer and its response is checked against the theoretical response.

Recalibration of the entire system is repeated twice a year and whenever the equipment is moved. Recalibration is also performed if any indication of faulty behavior is suspected.

Instrument Performance and Analysis of Instrument Response

To obtain the flow pore distribution of a membrane filter with the automatic flow pore analyzer, it is necessary only to have a sample of the membrane large enough to allow two 1-inch disks to be cut. One disk is wetted with a suitable fluid, such as kerosene, and the other is left dry. These filters are placed in filter holders and attached to the appropriate parts on the machine. The recording charts and air-pressure program valve are turned on. The machine then will print out the four functions: pressure, dry air flow, ratio of wet air flow/dry air flow, and the pore-distribution function.

The pressure curve relates the flow pore size to the ratio curve and also relates the air flow versus pressure for the membrane. The pressure response is printed on the flow pore distribution curve so as to align this curve to the proper pore size. Incorporation of a true x-y recorder would eliminate the need for the pressure scale being printed out for each test. The rate of pressure increase has been standardized so that distribution curves may be compared directly. With kerosene as the wetting fluid, the relation between the pore size and pressure, as previously stated, is

$$\text{Pore size in microns } (\mu) = \frac{12.5}{\text{pressure in psi}} \ .$$

The dry-air-flow curve is simply the air-flow rate passing through the dry filter under the differential pressure indicated by the pressure pen. If this curve were drawn on a rectilinear graph with the y axis as flow and the x axis as pressure, it could be seen that the air flow increases faster than a linear response to pressure. This result probably is due to the changing Reynolds number of the air passing through the pores of the filter. The factor is believed to have a slight effect upon the pore-distribution relation.

The curve representing the ratio of wet air flow to dry air flow is most important. The value of this ratio, of course, can vary only between 0 and 1, and, when expressed as percent, it defines the percentage of flow pores open to the specific pressure reading. For example: at a chart reading showing a ratio of 0.5, 50 percent of the flow pores are open and 50 percent are not opened. The corresponding pressure, as recorded on the polar-coordinate graph, divided into 12.5 (with kerosene as the wetting fluid) yields the mean flow pore for the filter. This is the same mean flow pore as defined by the ASTM standard D-2499. Any other percentage of the total porosity may also be directly found from this curve. The relative contribution to total flow rate of any pore size is shown on the flow pore-distribution diagram. This representation is actually the derivative of the ratio curve.

The flow pore distribution curve that is plotted on rectilinear chart paper shows the relative number of pores that exist at any pore size for a given filter. The curve does not provide absolute values nor does it show differences between filters in total numbers of pores. Differences in the shapes of the curves and the relative numbers of flow pores that exist at different sizes are evident. Figure 6 shows the comparison between 0.45-μ cellulose triacetate and cellulose nitrate membranes.

Membrane Interaction With Beer. Filtrations through membranes have shown clogging effects that are inexplicable in terms of the sieve-like action of filter associated with particulate separations. These effects do not include chemical attack and/or the swelling of the membrane filters. Membrane filters are in a pore-size range where it is not unlikely that interactions between the membrane and filtrate materials could occur. They could affect the efficiency of

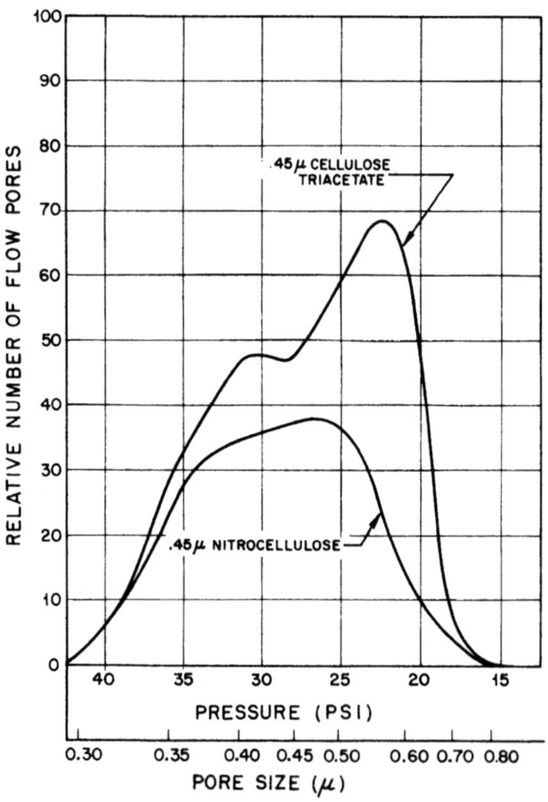

Figure 6. Comparative flow pore distributions.

the filtration. The clogging effects noted are ascribed to interactions on the surface of the membranes between the polymer and soluble components of the filtrate. When such particular attractions exist, significantly different membrane capacities may occur. Thus, two products of a given brewery were filtered through each of two membranes having approximately the same pore-size distribution. One of the filters was a cellulose ester and the other was an experimental PVC membrane. One of the beers was a light malt liquor. The other was a heavy beer. It was found that the heavy beer experienced a greater throughput through the cellulosic membrane than it did through the experimental PVC. The lighter beer, however, excelled in its throughput through the experimental membrane. Water containing colloidal ferric hydroxide found an even greater discrepancy in the relative membrane capacities before clogging. The cellulose ester membrane showed the more rapid clogging rate.

If ordinary clogging were involved, the two beers could have been expected to mirror the results found for the colloidal-iron oxide-water system. Instead, there was actually a reversal of membrane rank with the heavier beer. This occurrence is attributed to the interaction of the membrane surface with proteinaceous material of the beer. Interestingly, the clogging layer could be removed, without benefit of backwashing, by substituting the lighter beer for the heavier beer during the filtration operation, at which time the experimental PVC filter was restored to the throughput qualities of its pristine state.

It is seen that the evaluation of filter efficiency must take such interactive possibilities into account. A membrane material that is best for one beer need not be best for others. These interaction effects may be sufficiently dependent on the nature of the polymer of which the filter is constructed as to make the wrong choice of material an uneconomical one.

There is some trend in the brewing industry toward lighter, less filling beers. Beer-throughput comparison tests using commercial filters show that Dynel membranes have a substantially higher capacity to resist clogging than do equivalent nitrocellulose membranes for such light beers.

Light-Beer Filtration Through Commercial Membranes

Test	Dynel	Nitrocellulose
1	3.16	3.00
2	3.18	2.72
3	3.31	2.86
4	3.94	2.07
5	3.32	2.31
6	3.48	1.79
7	3.34	2.79
8	3.33	2.48
9	4.18	2.22
Average	3.47	2.47

Filtration results are given in total liters throughput per cm^2 of membrane surface. Filtrations were run at a constant flow rate of 100 $ml/min/cm^2$ with a cutoff pressure of 60 psig.

While the interaction of the filter surface with what is being filtered exerts an influence on the throughput efficiency of the membrane, the contribution of pore size still exerts its customary sway. Thus, in the filtration of an extra heavy beer, a series of Dynel membranes showed the expected correlation between mean-flow pore size and throughput volume.

Heavy Beer Throughput Results for a Range of Dynel Membranes

Mean-Flow Pore Size (μ)	Air-Flow Rate*	Beer Throughput**
3.6	33	21.75
3.3	31	9.21
3.0	30	13.15
2.6	30	5.18
2.4	25	4.75
2.3	24	4.50
2.0	22	4.04
1.9	20	2.90
1.6	17	2.90
1.4	13	2.18
1.3	15	1.53
1.2	14	1.19
1.1	8	0.73
0.91	8	0.56
0.55	5	0.13
0.47	4	0.06

*Air Flow Rate: Liter/min/cm^2 at 70 cm Hg differential pressure.

**Filtration results are given in total liters throughput per cm^2 of filter surface area in a constant-flow-rate filtration (30 ml/min/cm^2) with a 60 psig cutoff point.

The experimental PVC-type membrane, presumably as a result of interaction with the proteins of heavy beers, is less suited for their filtration. That an interaction with the membrane surface is involved is made evident, as is the contribution of the particular surface chemistry, by the alteration in filter efficiency that is occasioned by the coating of the experimental PVC membrane with a more hydrophilic polymer. A marked advantage to heavy-type beer filtration results from this treatment.

Comparison of Uncoated and Hydrophilic-Coated PVC-Type Membranes in Heavy-Beer Filtration

Beer A	
Experimental PVC	175
Hydrophilic Coated	263

Beer B	
Experimental PVC	241
Hydrophilic Coated	720

Filtration results are given in total ml throughput per cm^2 of filter surface. Filtrations were run at a constant-flow rate of 30 $ml/min/cm^2$, with a cutoff pressure of 60 psig.

Credence is given the surface-interaction concept by the salubrious results obtained by hydrophilic coating. If anything, the pores of the membrane were diminished by this treatment. Clearly then, the aggrandizement of the beer throughput was not due to particulate-filtration phenomena.

Membrane Interaction With Albumin and Serum. The interaction effects between membrane filters and beer are seen as involving protein. Clogging is attributed to the deposition of such protein upon the filter surfaces through the agency of attractive forces like hydrogen bonding. It is known that, when beers are brewed from grains that contribute albumin to the beverage, membrane filtration is especially prone to early clogging. However, when the albumin is insolubilized by "boiling" the beer, membrane filtration unimpeded by undue clogging becomes possible.

The qualitative ranking of membrane filters by the extent of their interaction with albumin is easily achieved. A standardized drop of an aqueous solution of serum albumin was placed upon the surface of each of three membrane filters. The extent to which the water-drop spread was evidenced by the area of wetness. This was essentially the same for the three filters of cellulose triacetate, cellulose nitrate, and experimental PVC. However, the spread of the albumin within the wet area differed. The visual detection of the albumin was made possible by the use of Ponceau S stain. It was found that the albumin spread along with the water over the cellulose triacetate when the bovine-albumin concentration was 100 mg percent. At that concentration level, the albumin spread only very little on the cellulose nitrate and even less on the experimental PVC. At a concentration of 700 mg percent, the spread of albumin over the experimental PVC filter did not increase much but did so for the cellulose nitrate. This experience is interpreted as indicating the relative binding forces for the three polymers for serum albumin in aqueous solution. The cellulose triacetate is seen as exerting the least interaction – hence, the least binding and the least impediment to spreading along the membrane surface. The experimental PVC by this measure binds the strongest of the three polymers examined. The nitrocellulose binds albumin strongly, but the attractive forces are saturated at the 700-mg-percent level, permitting the further spreading of the protein. The interaction strength of membranes is, thus, expressible in terms of its fixative properties, both in terms of the quantity of protein it can bind and in the extent of localization the bound protein exhibits.

A more pronounced difference in membrane-filter performance occurs when bovine serum is the filtration fluid. Comparison was made (see Table 1) between a 0.22-μ cellulose triacetate membrane and a 0.45-μ experimental PVC membrane as a filter for bovine serum. In order to minimize clogging due to particulate matter, the bovine serum was prefiltered through a 0.8-μ cellulose triacetate membrane. It was found that the cellulose triacetate membrane, although characterized by a smaller mean-pore size, gave double the throughput of the experimental PVC filter. The difference in performance is ascribed to filter interaction with serum protein.

The implications of such experiments to protein systems other than those directly measured are uncertain at best. Extrapolations must, therefore, be made with circumspection. Thus, in the albumin-fixing measurements, cellulose triacetate shows the least tendency to

TABLE 1. SALES SPECIFICATIONS

0.45-μ Dynel[(a)] and 0.20-μ Cellulose Triacetate Membranes

	Triacetate	Dynel
Mean flow pore, μ	0.20	0.45
Polymer	Cellulose Triacetate	PVC-acrylonitrile
Autoclavability	Yes	Yes[3]
Maximum temperature, C	150	125
Surface	Smooth	Pattern[4]
Air-flow-rate[1]	3.9	6
Water-flow rate[2]	30	7
Thickness, μ	140	140
Kerosene bubble point, psig	24	12
Water bubble point, psig	50	15

(a) Dynel is a registered trademark of Union Carbide.

Notes

(1) Air-flow rate in liters/min/cm^2 at a differential pressure of 70 cm Hg.
(2) Water-flow rate in ml/min/cm^2 at a differential pressure of 70 cm Hg.
(3) Autoclavable in filter holder.
(4) Pattern due to the presence of woven nylon substrate.

interact with the protein, experimental PVC interacts the most, and nitrocellulose is in between. The experience with beer filtration utilizing these membranes shows a different interaction order.

Comparison of Heavy-Beer Throughput for 0.8-μ Nitrocellulose, Dynel, and Cellulose Triacetate Membranes

	Nitrocellulose	Dynel	Cellulose Triacetate
	1.44	1.05	0.19
	1.55	1.25	0.24
	1.18	0.89	0.22
	1.30	1.23	0.26
	1.24	.99	0.18
	1.51	.74	0.16
Average	1.37	1.03	0.21

Filtration results are given in total liters throughput per cm^2 of filter surface area in a constant-flow-rate (30 ml/min/cm^2) filtration with a 60-psig cutoff point.

Protein systems are notoriously complex in terms of molecular species, the influence of pH on charge distributions, the possibilities for conformational alterations, etc. Their behavior in filtration operations are, therefore, often difficult to predict. For this reason, the suitability of a membrane filter for a particular operation involving proteins merits experimental assessment.

The various techniques for measuring porosity offer a way of disclosing the morphological structure of a membrane that flow measurements alone cannot reveal. For example, if the density of a given filter is high, it may indicate that a degree of compaction of the membrane has occurred. The static flow measurements may not have been materially affected, but the membrane may show a rapid clog rate during filtration runs. These evaluations are of prime interest to the membrane manufacturer in establishing effective process control.

Pore Size and Distribution. There are in existence several methods for measuring the pore sizes in a membrane filter. Among these are electron microscopy, mercury intrusion, and the mean-flow pore method described above. It has been our experience that any statement of membrane-filter pore-size definition should be accompanied by a reference to the method used. The various measurement techniques do not necessarily yield the same values. Pore size is not the best indication of retention capability or of clog characteristics. In the case of bacterial retention, pore-size values often seem inconsistent with filtration test results. Typically, a membrane with a maximum pore size of 1.0 μ will retain bacteria of 0.5-μ size. Additionally, membranes can be prepared with a 1.0-μ maximum pore size that will not retain 5.0-μ bacteria. Two membranes made from the same polymer and having the same mean pore size may exhibit very different clog characteristics. An extension of the concept of pore size is needed to explain these experiences. Previously, pore-size distributions were difficult to obtain. Hence, all too many conclusions were inadequately based on mean pore-size measurements alone. A possible assistance is offered by the previously described instrument. It is capable of providing an automatic printout of flow pore distributions. Experiments with this equipment have shown that, while membranes with equal mean flow pores may exhibit the same flow-rate behavior, further analysis of the complete flow pore distribution is necessary to explain comparative retention and clog capabilities.

Figure 7 shows the comparative flow pore distributions of two different membrane filters that have equal flow porosities and equal mean flow pores and thus exhibit the same flow-rate properties with clean fluids. (Note: mean flow pore is by definition represented at that graphical point, A$'$, which, when extended back to the pore-size axis, divides the area under the distribution curve in half. Total flow porosity is represented by the distribution curve when the curve is the result of a differentiation of the wet flow to pressure ratio with respect to pressure.) The membrane represented by curve B can be expected to exhibit poorer retention of 5.0-μ bacteria than that represented by curve A. This difference occurs as a result of B's comparatively greater numbers of flow pores in the larger-than-0.5-μ region.

In the filtrations of particles of a given size, we have seen strong indications that the most efficient filter from the standpoint of clog resistance is one which has its flow porosity concentrated at a size somewhat smaller than that of the particles. Filters that have a comparatively large percentage of their flow porosity concentrated at the size of the particles they are filtering tend to clog rather quickly. We hypothesize that the one case is similar to marbles being retained by window screen, wherein "clogging" would be minimal. The other case

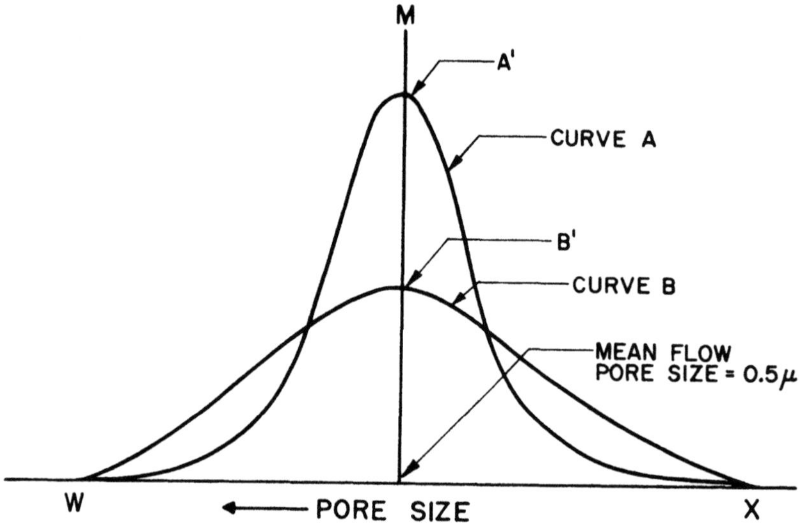

Figure 7. Hypothetical flow pore distribution.

involves the particle fitting more precisely into and over the pores. Clogging would then be enhanced. Figure 8 shows the pore distributions of two membranes that have the same flow porosity, the same mean pores, and the same bubble points. Experimental evidence shows that, in the filtration of 0.7-μ particles, the filter represented by curve C will clog more slowly than will the filter represented by curve D. This is explained by the comparatively large number of flow pores at or near 0.7 μ in membrane D. Referring back to Figure 7, a filtration of 0.5-μ

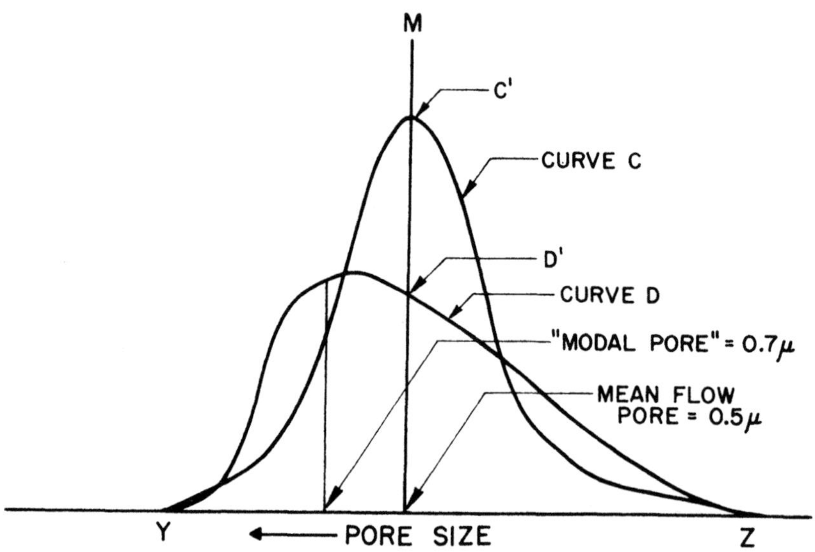

Figure 8. Hypothetical flow pore distribution.

particles indicates, though less conclusively, that the membrane represented by curve B clogs more slowly than does that represented by A. This is probably so because of the difference in relative flow pore numbers at the 0.5-μ level, which, coincidentally, is the mean flow pore for both membranes in this case.

Pore-Shape Contributions. The influence of pore shape on membrane filtration performance was noted. Two different types of membranes, both of cellulose triacetate, were used in this study. Type I had customary sponge-like pores that uniformly characterized it throughout its bulk. Type II membranes were characterized by a skin effect on one surface. This produced an asymmetric membrane (see Figure 9) wherein the pore structure at one surface differed from that at the other. Both Type I and Type II membranes were matched with regard to air-flow rate, mean flow pore size, and thickness, as well as in being made from the same polymer. However, they showed differences in clogging rates when used to filter bovine serum. Both being made of cellulose triacetate, their interaction with the serum protein was considered equal. Their clogging therefore was attributed to the action of particulate matter present in the serum.

Three effects were noted in filtration experience:

(1) No side differences were found in filtrations involving Type I membrane.

(2) In filtrations involving Type II membranes, skin-side downstream filtration showed far greater throughput capability than did skin-side upstream filtration.

(3) Skin-side upstream filtrations with Type II filters showed much poorer throughput characteristics than did bovine-serum filtration involving Type I membranes.

Clog Characteristics of Skinned vs Unskinned 0.2-μ Cellulose Triacetate Membranes With Bovine Serum

Type I Membrane (Unskinned)	Average Throughput, ml/cm^2
Skin side upstream	40
Skin side downstream	41
Type II Membranes (Skinned)	
"Top" side upstream	30
"Top" side downstream	40

These experimental findings are rationalized as follows: the Type I membrane, being regular in porosity throughout its bulk, showed no anisotropic effects during filtrations from either side. Type II membranes did show a side-to-side difference in filtration capability that was a reflection of their asymmetry. Type II membranes are believed to exhibit "funneling" characteristics in that the skin side has far fewer openings than does the opposite side of the membrane. This accounts for the difference in clogging rate from side to side in the Type II membrane. When any of the fewer openings on the skin become blocked, the effect in terms of

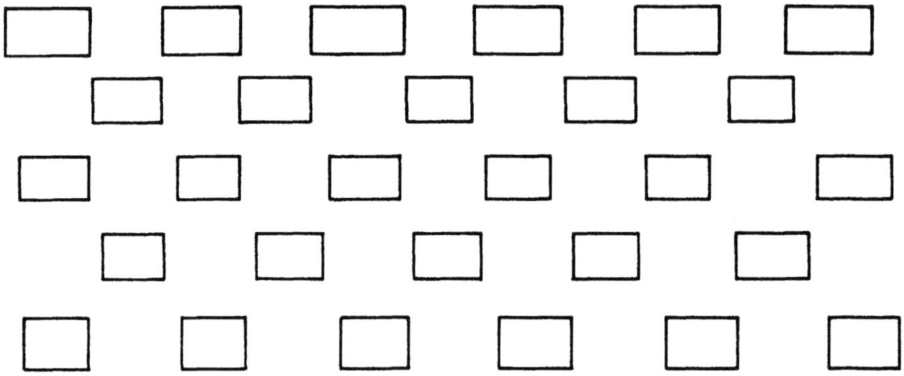

SCHEMATIC TYPE I MEMBRANE

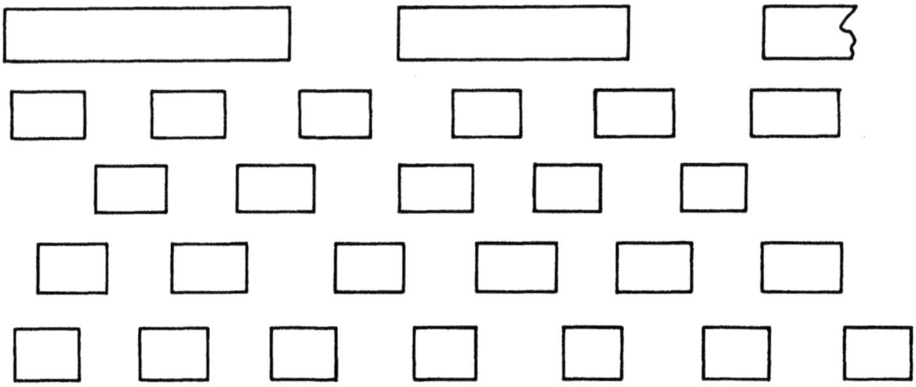

SCHEMATIC TYPE II MEMBRANE

Figure 9. Skinned versus unskinned membrane.

volume throughput become profound. Pore blockages on the other side had a lesser effect. Hence, Type II showed a higher clogging rate when its skin side was used upstream in bovine-serum filtrations.

It can be seen from the foregoing that membrane filtrations are phenomena that involve the input of many membrane parameters in a complex manner.

Computer Analysis. The membrane-filter qualities of filtration rates, clog characteristics, and retention capabilities depend upon an interrelationship of many factors. The elucidation of the respective influences of each of these, for instance, on the throughput volume in beer filtration can hardly be assessed by single-parameter variation studies. A computerized investigation is needed to handle simultaneously the large number of variables concerned.

What is sought in any such study is the causality between a final effect, such as beer throughput volume, and the factors that control it, such as membrane porosity. Initially, these parameters are selected rather intuitively. When their collective influence is assessed as having a definite relationship to the end effect, causality (though not proven statistically) is considered demonstrated. Otherwise, one refines the analysis to include other factors. This continues until the relationship between cause and effect is sufficiently established.

This analytical technique itself requires computer techniques as an aid in deriving prediction equations. The statistical methods in the process of adaptation are regression and correlation analysis. Regression analysis refers to a group of methods used to predict the value of one dependent variable (generally referred to as the "Y" variable) from different experimentally obtained values of one or more independent variables (generally referred to as "X_1", "X_2", "X_3", etc.). Simple regression involves the use of one independent variable and derives a least-squares line equation of the form

$$Y = AX + \text{constant}.$$

Multiple regression involves the use of two or more independent variables and may derive an equation (for three independent variables) of the form:

$$Y = AX_1 + BX_2 + CX_3 + DX_1X_2 + EX_1X_3 + FX_2X_3 + GX_1X_2X_3 + \text{constant}.$$

However, most computer programs do not generate the interaction factors. The usual regression equation corresponding to the above example is of the form

$$Y = AX_1 + BX_2 + CX_3 + \text{constant}.$$

The correlation coefficients describe the closeness of the relationship between the dependent variable and the independent variable or variables. If the relationship is one of proportionality, which is what is being sought, the coefficient is +1. If the value of the dependent variable is random with respect to the value of the independent variable or variables, the coefficient is zero. If the relationship is one of inverse proportionality, the coefficient is -1.

One series of computer analyses has been run using the data from sixteen different experiments. The regression equation was derived for the purpose of predicting beer-throughput capability for a group of commercially available, nylon-supported Dynel-membrane filters ranging in mean flow pore size from 0.47 to 3.60 μ. The regression was run as a log transform:

Independent Variables

X_1 = Mean flow pore

X_2 = Air-flow rate (liters/min/2.9 cm^2 at 70-cm-Hg differential pressure. The effective filtration area of 1-inch disk is 2.9 cm^2.)

Dependent Variable

Y = Beer throughput (ml/cm^2 through 60 psig cutoff)

Regression Equation

$$1.9349 \, \mathrm{Log}_{10} X_1 + 0.5988 \, \mathrm{Log}_{10} X_2 + 1.9404 - \mathrm{Log}_{10} Y$$

or

$$87.18 \, (X_1^{1.9349}) \, (X_2^{0.5988}) = Y.$$

The multiple correlation coefficient was found to be +0.9861. The regression equation thus proved to be very useful in predicting comparative filtration behavior of membranes with large differences in mean flow pore size and air-flow rate. However, prediction of filtration differences between membranes closely related in mean flow pore size and air-flow rate proved inadequate. This result indicated to us that, while some of the important variables necessary for the desired prediction had been considered, not all the operative parameters had been taken into account.

A more complex series of computer analyses is planned. The independent variables to be regressed are mean flow pore, air-flow rate, polymer type, and relative flow porosity. The results of this work are expected to yield an improved prediction equation. Ultimate predictability, we believe, awaits, in part, the developing of an adequate mathematical analysis of the flow pore distribution curve, of which "relative flow porosity" is but a simplistic approximation. We hypothesize that Fourier analysis or other curve-fitting techniques will be helpful in this regard.

SEPARATION OF BLOOD SERUM PROTEINS BY ULTRAFILTRATION

Carel J. van Oss and Paul M. Bronson
Department of Microbiology
School of Medicine
State University of New York at Buffalo

INTRODUCTION

The majority of human blood serum (non lipo-) proteins exists in only three discrete sizes: roughly 70 percent are of the size of serum albumin, with a molecular weight around 70,000 (sedimentation coefficient 4 S); some 25 percent, the globulins, have a molecular weight of 160,000 (7 S); and about 5 percent, the macroglobulins, have a molecular weight of 860,000 (19 S)[1].

There are also small amounts of microglobulins (smaller than albumin), some globulins intermediate in size between globulins and macroglobulins (M=300,000; 10 S), and some supermacroglobulins of a molecular weight of several millions (23 S, 28 S, 33 S, etc.)[1].

The first membrane to be treated here is one that stops all normal serum proteins. This membrane is of a crucial utility, because it allows the rapid reconcentration of all serum protein fractions that have become excessively diluted through any prior separation or purification treatment[2], be it preparative electrophoresis, ion exchange or gel filtration chromatography, $(NH_4)_2SO_4$ precipitation and dialysis, density gradient ultracentrifugation, or fractionation by ultrafiltration.

We shall then discuss membranes that separate the main size groups of blood serum proteins, and we shall finally describe some separations of a few of the minor size groups of proteins by ultrafiltration.

Full descriptions of the preparation of a number of anisotropic ("skinned") cellulose ester membranes have been given elsewhere, and will be repeated here only in outline, but with enough detail to allow reduplication of the membranes.

A MEMBRANE THAT RETAINS ALL HUMAN BLOOD SERUM PROTEINS

Membranes that stop all proteins have been routinely available, since at least the late 1940's, in the form of cellophane sheets and tubing. These (homogeneous) regenerated cellulose membranes are quite reproducible, have a uniform pore size almost exactly calculated to stop all blood serum proteins, are generally available in large quantities, and are quite cheap. Their sole drawback is the flow rate that can be attained with them, which is about 3 ml/hour/100 cm^2 membrane/30 psi, when ultrafiltering a 0.7 percent protein solution. Nevertheless, with heroic measures such as the use of pressures of from 300 to 500 psi and of membrane surface areas of several square feet, appreciable quantities of ultrafiltrate (of the order of liter per hour) can still be obtained with tubular ultrafilters[(3)]. But the necessity of working with large surface areas of membrane imposes a rather large minimum volume on the ultrafilter. Its yield is attractive enough when its main use is to obtain large volumes of ultrafiltrate. But if one aims at concentrating a protein solution, the large minimum interior volume (of about a pint of liquid per square foot of membrane), imposes a minimum of about 2 gallons on the volume of solution that can be concentrated, say, 10 times.

The discovery by Loeb of anisotropic ("skinned") membranes in 1962[(4)] was a great advance because, although these membranes were initially mainly intended to retain salt by "reverse osmosis", their principle was soon extended to more porous membranes which could ultrafilter the solvent of diluted protein solutions at flow rates of several hundred ml/hour/100 cm^2 membrane/30 psi[(5,6,7)]. Curiously enough, the high pressure tubular system described above is now still among the more suitable "reverse osmosis" systems for water desalination, where drastic concentration of small volumes is less of a requirement than high pressure and a high flow rate of ultrafiltrate in the form of fresh water[(8)].

Very high flux anisotropic cellulose acetate membranes can now be made which retain all normal serum blood proteins. A membrane of this type makes the concentration of about a pint of a dilute protein solution down to only a few milliliters a matter of only a few hours, without any protein denaturation. The use of such a membrane in the clinical laboratory has been recently described[(7)]. A more detailed description of various physical properties of this membrane has also been given recently[(7a)] and only the major properties of the membrane will be given here. It was found that the optimal pressure for concentrating proteins with this membrane is close to 30 psi. The optimal thickness of the membrane is 0.15 mm. As seen by scanning electron microscopy the thickness of the actual protein stopping skin is of the order of 1 micron. The scanning electron micrographs in Figure 1 illustrate quite clearly the difference in texture between the top of the membrane and the very porous coarse bottom of the membrane. With the detailed description of the properties of this membrane, a comparison was also made between the membrane and a few of the commercially available membranes[(7a)]. The membrane always has the highest flow rate when it has never been dried and is conserved in cold water from its inception on. Nevertheless drying of the membrane is possible, with the help of either a prior impregnation in 50 percent glycerol or in 0.1 percent sodium dodecyl sulfate, but the flux of the membrane is reduced by about 20 to 25 percent after drying. A description follows of the preparation of the membrane:

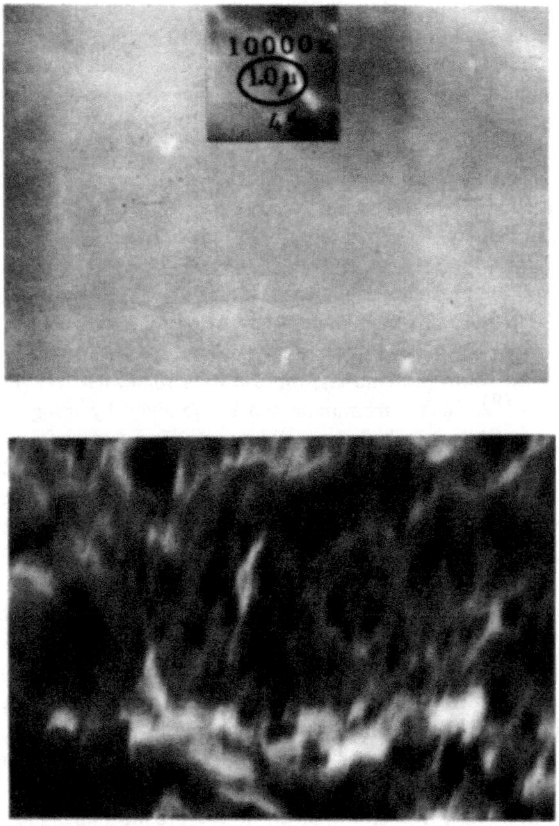

Figure 1. Scanning electron micrograph of the protein-retaining membrane. Top: "skinned" top of the membrane (approximately 1 μ thick); bottom: porous bottom of the membrane. X10,000.

Cellulose diacetate (39.8 percent acetate, ASTM viscosity 3, Eastman No. 4644), kept in a vacuum desiccator, gives the best results. A 25 g amount of this is added slowly, to a mixture of 75 ml of acetone and 50 ml of formamide in a mortar, under constant stirring with the pestle until all lumps are dissolved. The mixture is then poured into a 500 ml filtering flask which is stoppered and placed in a water bath kept at 50-55 C. Vacuum is applied (approximately 40 cm Hg) to the flask until all air bubbles have disappeared. Approximately 15 ml of the mixture is then poured in a ribbon across the short side of a 17 x 25 cm glass plate (along the two long edges of which two 0.15-mm thick metal runners are clamped). The mixture is then spread quickly and evenly over the glass plate, with the help of a 15-cm long glass test tube, by drawing the tube in one smooth horizontal motion along the runners. An excess of about two thirds of the applied mixture is swept off the glass plate and discarded. As quickly as possible, the glass plate with the spread-out mixture is submerged in an ice-water bath. It is kept there for at least 1 hr, after which time the membrane can be lifted from the plate. The almost instantaneous coagulation of the upper surface of the mixture, upon its immersion in cold water, causes the membrane to be provided with a very thin skin which is denser than the rest of the membrane structure. This is the actual protein retaining skin, and care must be taken always to use the membrane with the side up that was away from the glass plate when it was formed. For immediate and even for future use, the membrane is best kept in cold (tap) water. In order not to lose sight of the skinned side of the membrane, the best practice is to cut circular membranes out of the larger sheet only as needed, and to mark the edge of the larger sheets in some asymmetrical fashion(7).

When retaining protein from a 1 percent solution, the flux of this membrane is about 300 ml/hour/100 cm^2/30 psi. As this membrane tends to pass about 5 percent of the proteins of a molecular weight of 23,000 (when there are any present, as for instance in the serum or urine of patients with Bence Jones proteinuria), the use of a somewhat less porous membrane may sometimes be desirable(8). Such a membrane can be obtained by using only 35 ml formamide per batch in the preparation described above. When retaining protein from a 1-percent solution, the flux of this tighter membrane is about 160 ml/hour/100 cm^2/30 psi. On a laboratory scale, the best results are obtained with these membranes in an ultrafilter that allows for constant stirring of the protein solution to avoid clogging of the membrane by a concentrated layer of the protein that is being retained.

High flux protein-retaining membranes are quickly developing into the work horses of all biochemical and chemical laboratories dealing with proteins. A variety of separations that were hitherto theoretically possible but simply not practicable, because the tremendous dilution of the proteins they entailed, can now be undertaken routinely because after any manipulation resulting in vast dilution it has become a very simple matter to reconcentrate the protein quite rapidly without any trace of denaturation (Figures 2 and 3) in an ultrafilter, using either the membranes described above or one of the commercially available membranes [the Diaflo UM 10 membrane from Amicon Co., Lexington, Mass., or the P.E.M. membrane from Gelman Instrument Co., Ann Arbor, Mich.; the membranes marketed by the Millipore Co. and by the Sartorius Co. were found to leak protein consistently(8)].

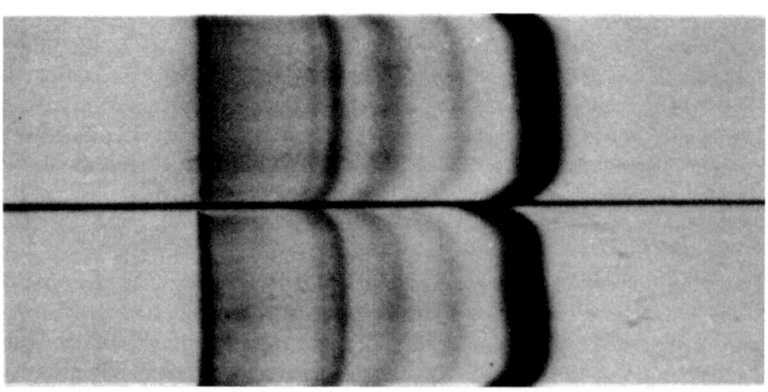

Figure 2. Cellulose acetate electropherogram of normal human serum (top) and of the same serum 400 x diluted with saline and reconcentrated to its original protein concentration by ultrafiltration (bottom).

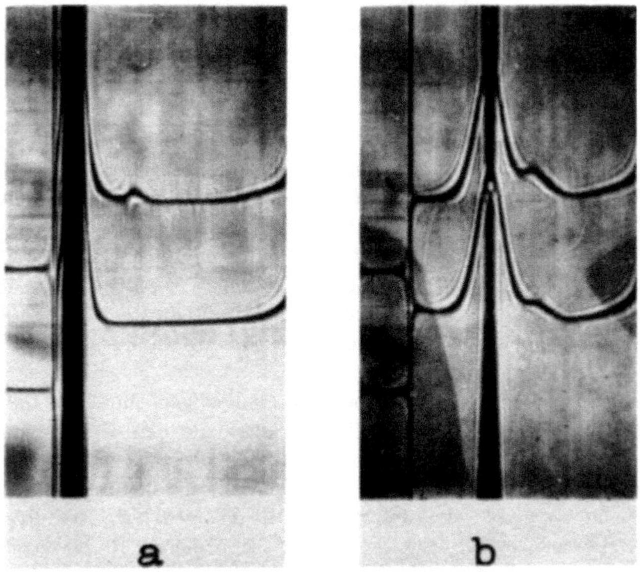

Figure 3. Analytical ultracentrifugation patterns of normal human serum (top curve) and of the same serum 400 x diluted with saline and reconcentrated to its original protein concentration by ultrafiltration (bottom curve). The protein concentration was adjusted to 1.0 percent. The left picture was taken 12 minutes and the right picture 44 minutes after attaining 59,780 rpm. The schlieren angle was 60 degrees.

MEMBRANES THAT RETAIN ALL MACROGLOBULINS

There are two sizes of immunoglobulins (antibodies), the majority having a molecular weight of approximately 160,000 and a small but important minority, the macroglobulins, having a molecular weight of approximately 860,000. These two different types of immunoglobulins tend to be synthesized at different stages of antibody formation. For that reason it is of great utility to be able to separate them and to test them for antibody activity in general immunology as well as in immunohaematology. There are other physical methods for separating the macroglobulins from the smaller immunoglobulins and there are also chemical methods for destroying the antibody activity of the macroglobulins only[9]. However, these methods tend to be rather laborious. Moreover, with the physical methods a certain degree of overlapping of the fraction is hard to avoid, and with the chemical methods the dosage at which the macroglobulin activity is completely destroyed is difficult to differentiate from the dosage at which most of the lower molecular weight antibody activity is inactivated. In an attempt to improve this state of affairs, we designed, a few years ago[10], an anisotropic membrane made of agarose, which can be cast in a Gooch funnel and which, upon the application of a vacuum beneath it, will ultrafilter 1-2 ml of serum in about 6 hours and which passes only all those serum proteins that have a molecular weight under about 300,000.

The extreme slowness of this membrane, although it is "skinned", is due to its great thickness (2-4 mm). But now we have developed(11) a thin, skinned membrane of a mixture of cellulose nitrate and cellulose acetate which will yield the same result in a matter of minutes.

The membrane is made as follows:

> A 500 ml Serval Omnimixer chamber is charged with 50 ml of glacial acetic acid and 40 ml of acetone. To this mixture is added 15 g of cellulose nitrate (30 percent ethyl alcohol, viscosity 16.2, A.B., Grade DHB 14 E, DuPont). The whole is then mixed using the Omnimixer, and once the mixture is uniform, 15 g of cellulose diacetate (39.8 percent, ASTM 3, Eastman No. 4644) and 40 ml of formamide are added to the chamber. Mixing is resumed until the mixture is uniform. From this point, the process continues as previously described.
>
> The mixture is cast in the same way as described above for the protein-retaining membrane. Here, 0.25-mm thick runners are used, while stainless steel plates have to be used instead of glass plates, to prevent the membranes from sticking to the plates. Once drawn, the membrane is "superskinned" with the aid of an air blast [at 20 C; see also Reference (10)] provided by a hair dryer (Oster, Model 202, Milwaukee, Wis.) held 15 cm above the membrane for 60 seconds. Then the membrane is immersed in an ice water bath in which it remains for at least 1 hour prior to use. This membrane can be dried after immersion in 50 percent glycerol but, due to its strong anisotropy (see Figure 4) and to its brittleness in the total absence of plasticizer, it cannot usefully be dried after pretreatment with 0.1 percent sodium dodecyl sulfate.

With this membrane the optimal pressure is 10 psi, but pressures up to 20 psi can be used. The actual "skin" is about 0.3μ thick as judged by scanning electron microscopy. Figure 4 illustrates the difference in texture between the top and the bottom of this membrane.

The flow rates with this membrane are about 150 ml/hour/100 cm^2/10 psi for 1 percent protein and 60 ml/hour when ultrafiltering whole serum. In practice it is rarely feasible to obtain more than 10-20 percent of the protein in the ultrafiltrate, as the protein concentration of the ultrafiltrate is significantly lower than that of the solution being ultrafiltered. (For example, when whole serum with 7.5 percent protein is being ultrafiltered, the ultrafiltrate will contain 1.8 percent protein). But the important aspect is that the ultrafiltrate will be devoid of macroglobulins, while albumin and the 7 S globulins are present in close to normal proportions. Figure 5 shows the ultracentrifugal analysis of a serum, before and after removal of the macroglobulins by ultrafiltration. Immunodiffusion tests (not illustrated herein) have confirmed the retention of 19 S IgM macroglobulin (M=860,000) and the passage of 7 S IgG globulin as well as of 7 S IgA globulins (M=160,000-180,000).

The "cut-off" of this membrane is at M=300,000. This was found when a preparation of human IgG which contained about 15 percent of IgG dimers (10 S, M=320,000) was ultra-filtered. The dimers were retained, but the membrane got completely clogged after a period of time (which does not occur in the absence of proteins of that size). This behavior tends to indicate that the molecular size of that protein is just about of the same order of magnitude as the pore size of the membrane. Without agitation the membrane gets clogged almost immediately when a 10 S globulin-containing solution is ultrafiltered and the small volume of

Figure 4. Scanning electron micrograph of the macroglobulin-retaining membrane. Top: "skinned" top of the membrane (approximately 0.3 μ thick); bottom: porous bottom of the membrane. X 3,000.

ultrafiltrate obtained is then practically devoid of protein. But with adequate agitation, the membrane can be used for the removal of 10 S from human IgG preparations.

For work on blood group antibodies(12), the membrane can also be used in a simple centrifuge tube/ultrafilter cell, as made by the Millipore Corporation, Bedford, Massachusetts. Through the absence of stirring, the yield is smaller than with a stirred ultrafilter cell, but no specialized equipment (such as compressed nitrogen cylinders, valves, pressure gauges, ultrafilter cells) is needed, so that these membranes can be used in any laboratory equipped with a table top centrifuge. Typically, 3-5 ml batches of serum are centrifuged at 2500 g for 20 minutes, yielding about 1 ml of ultrafiltrate with 1.5 percent protein devoid of macroglobulins.

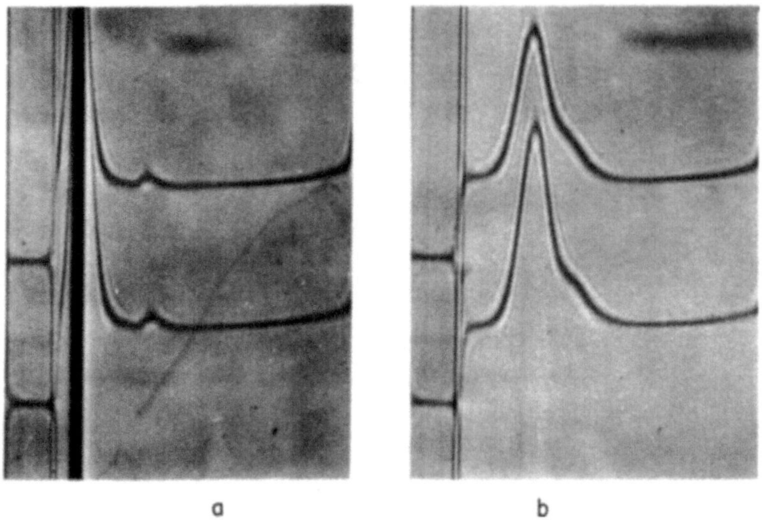

Figure 5. Analytical ultracentrifugation patterns of normal human serum (top curve) and of the same serum ultrafiltered through a macroglobulin-retaining membrane (bottom curve). The protein concentration was adjusted to 1.0 percent. The left picture was taken 12 minutes and the right picture 60 minutes after attaining 59,780 rpm. The schlieren angle was 60 degrees.

MEMBRANES PASSING ALBUMIN BUT RETAINING GLOBULINS

Membranes of this class are now commercially available from the Amicon Corporation (Lexington, Mass.), as Diaflo XM100. According to the manufacturer these membranes will stop globulin and pass albumin when the total protein concentration is 0.1 percent or less. In our hands, using a mixture containing 0.05 percent human serum albumin and 0.05 percent human serum gamma globulin, no separation of these two proteins by the XM100 membrane could be discerned when the reconcentrated (by ultrafiltration) ultrafiltrates were analyzed in the analytical ultracentrifuge.

On the other hand, an anisotropic cellulose acetate membrane, made in exactly the same way as our protein retaining membrane (see above), but with 100 ml of formamide per batch, would pass human serum albumin (M=69,000) and retain human gamma globulin (M=160,000) from the 0.05/0.05-percent mixture.

The disturbing aspect of the experiments with this membrane was that, when whole serum was diluted down to 0.1 percent total protein, no separation between albumin and globulins could be achieved. We suspected residual lipoproteins to play a role in this phenomenon, but thorough delipidation was of no avail. To this date a separation between albumin and globulins could only be obtained with this membrane when artificial mixtures were used consisting of 0.05 percent pure albumin and 0.05 percent 7 S gamma globulin (which contained 15 percent of 10 S dimers), and no explanation has as yet offered itself for the deviating behavior of diluted whole serum. (Although the fact that a separation between albumin and gamma globulin was only achieved when significant amounts of a gamma globulin dimer were also present, this may at least raise the suspicion that clogging of the membrane's pores with that dimer could have played a role in establishing the membrane's selectivity).

The fact that it has been possible to make a membrane (see above) which not only retains all macroglobulins but retains the dimer of IgG globulin (M=320,000) while passing IgG (M=160,000) itself, proves that it is possible to make a clean separation between two proteins of molecular weights 1M and 2M. This incites us to continue our search for a different and better membrane capable of separating albumin from the globulins, even when both are present at concentrations that are considerably higher than 0.05 percent, possibly even in whole serum.

MEMBRANES CAPABLE OF PASSING ALL OR MOST SERUM PROTEINS BUT OF STOPPING ALL VIRUSES

Membranes have been used for many years for the ultrafiltration of viruses, principally in order to determine their size [see, for instance, Elford's work on graded "Gradocol" membranes(13)]. But the use of membranes for analytical purposes has been superseded by better and more precise methods such as analytical ultracentrifugation and electron microscopy(14,15). It is not easy to determine from the literature whether these "Gradocol" membranes were homogenous or cryptoanisotropic(2). But no work has been published to date on the ultrafiltration of viruses with anisotropic membranes for preparative purposes.

The fact that it has proved possible to make a membrane (see above) that retains all macroglobulins, has incited us to try to make a slightly more porous membrane, capable of passing all (or most) normal blood serum proteins, while stopping all macromolecules of the size of the smallest known pathogenic viruses(14,16) which have a diameter of 18 mμ(15) which, for spherical particles, corresponds to a "molecular weight" of 2,400,000, or a sedimentation constant of 60 Svedbergs. To be absolutely sure of stopping all viruses one can of course use the macroglobulin-retaining membrane described above, which has a "cut-off" of about M=300,000. But, for practical reasons, a somewhat larger pore-size may be preferable. We have therefore aimed at making a membrane that will pass 19 S macroglobulins, while retaining all macroglobulins of a sedimentation coefficient of 27 S and over. Such macroglobulins are frequently found in the sera of patients with Waldenström's macroglobulinemia(9,17), so that these sera can conveniently be used for testing such membranes. We made such a membrane, in exactly the same fashion as the 19 S macroglobulin-retaining membrane, described above, but with 7.5 grams of cellulose nitrate instead of 15 grams. When 27 S macroglobulin containing serum of a patient with Waldenström's macroglobulinemia was ultrafiltered through this membrane, it retained all of the 27 S macroglobulin while it passed some of the 19 S macroglobulin. Thus, in all probability this membrane has the optimal pore size for retaining all known pathogenic viruses, while passing all, or most, normal blood serum proteins. For ultrafiltering plasma with the specific aim only of stopping serum hepatitis virus (which has a diameter of approximately 20 mμ(18,19), corresponding to about 90 S) the design of a more porous and presumably even faster membrane should be possible.

SUMMARY

It is now becoming evident that all major size groups of blood serum proteins can be fractionated by ultrafiltration on anisotropic membranes. Such membranes can be easily and reproducibly prepared in the laboratory.

Membranes that stop all serum proteins from the size of Bence-Jones (BJ) protein on (M=23,000), as well as very high flux membranes that stop all proteins of a size between BJ protein and serum albumin (M=35,000), are advantageously cast of cellulose acetate from an acetone/formamide solution. With the same ingredients a somewhat more open membrane can be made that will separate serum globulins (M=160,000) from albumin (M=69,000) in dilute solutions.

A strongly anisotropic membrane made of a mixture of cellulose acetate and cellulose nitrate will stop macroglobulins (M=860,000), but will pass regular globulins (M=160,000) and albumin at quite reasonable fluxes. A somewhat more open membrane of the same type will pass all serum proteins, while still having a sufficiently small pore size to stop all known viruses.

ACKNOWLEDGMENTS

This investigation was supported by Public Health Service Research Grant GM 16256 from the National Institutes of Health. The authors are deeply indebted to the technical assistance of Mrs. Cassandra Hayes, to Dr. James F. Mohn, Director of the Blood Group Research Unit of this Department for providing us continually with fresh serum, and to Miss Lynn E. Verhoeven of Alpha Research and Development Company, Blue Island, Illinois, for her expert help in making the Scanning Electron Micrographs.

REFERENCES

(1) H. E. Schultze and J. F. Heremans, *Molecular Biology of Human Proteins,* Vol. I, Elsevier, New York (1966).

(2) C. J. van Oss, in: *Progress in Separation and Purification,* (E. S. Perry, ed.), Vol. I, John Wiley & Sons/Interscience, New York (1968).

(3) C. J. van Oss, *Ultrafiltration.* Dissertation, Paris, 1955; J. H. de Bussy Publishing Co., Amsterdam (1955).

(4) S. Loeb and S. Sourirajan, *Advan. Chem. Ser.,* **38**: 117 (1962).
S. Loeb and F. Milstein, *Dechema Monographien,* **47**: 707 (1962).

(5) A. S. Michaels, *Ind. Eng. Chem.,* **57**: 32 (1965).

(6) W. F. Blatt, M. P. Feinberg, H. B. Hopfenberg, and C. A. Saravis, *Science,* **150**: 224 (1965).

(7) C. J. van Oss, C. R. McConnell, R. K. Tompkins, and P. M. Bronson, *Clin. Chem.,* **15**: 699 (1969).

(7a) C. J. van Oss and P. M. Bronson, *Separation Science* (submitted, 1969).

(8) S. Loeb, *Desalination,* **1**: 35 (1966).

(9) C. J. van Oss, in: *Advances in Immunohematology* (T. J. Greenwalt, ed.), Lippincott, Philadelphia (1967).

(10) C. J. van Oss, J. E. Lord, and A. Scheinman, *Nature,* **215**: 639 (1967).

(11) C. J. van Oss and P. M. Bronson, in preparation.

(12) C. J. van Oss, R. M. Lambert, J. P. Downing, and P. M. Bronson, in preparation.

(13) W. J. Elford, *Proc. Roy. Soc.,* Lond., B **112**: 384 (1933).

(14) B. D. Davis, R. Dulbecco, H. N. Eisen, H. S. Ginsberg, and W. B. Wood, *Microbiology* Part IV, Virology), Harper & Row, New York, 1967.

(15) R. M. Jamison, H. D. Mayor, and J. L. Melnick, *Exptl. Molec. Pathol.,* **2**: 188 (1963).

(16) F. L. Schaffer and C. E. Schwerdt, in: *Viral and Rickettsial Infections of Man* (F. L. Horsfall and I. Tamm, eds.), Lippincott, Philadelphia (1965).

(17) J. G. Waldenström, *Monoclonal and Polyclonal Hypergammaglobulinemia,* Vanderbilt University Press, Nashville (1968).

(18) M. E. Bayer, B. S. Blumberg, and B. Werner, *Nature,* **218**, 1057 (1968).

(19) I. Millman, V. Zavatone, B.J.S. Gerstley, and B. S. Blumberg, *Nature,* **222**, 181 (1969).

PRODUCTION OF ACIDIC SALT WITH SUBSTITUTION REACTION BY MEANS OF ION-EXCHANGE MEMBRANE ELECTRODIALYSIS

T. Nishiwaki, H. Hani, and S. Itoi
Asahi Glass Company, Ltd.

INTRODUCTION

Although ion-exchange membranes are utilized mainly for electrodialysis and dialysis on an industrial scale, they are finding practical applications at an increasing rate in other membrane processes, e.g., reverse osmosis.

Typical applications of ion-exchange membranes, as listed in Table 1, are of two kinds: (1) the separation of electrolytes from solution as in the desalination of saline water and (2) the separation of either nonelectrolytes from electrolytes or strong electrolytes from weak electrolytes. Furthermore, the various applications shown do not involve chemical changes, although physical changes, e.g., changes in concentration and composition, do occur.

Utilization of ion-exchange membranes as the diaphragms of dialyzers has made possible the separation of strong acids and salts that could not be achieved dialytically in the past. Today, these processes find remarkable applications in industrial fields, and they have been used by several nickel refineries for the recovery of sulfuric acid.(1)

Since being introduced over 15 years ago, ion-exchange membranes have been manufactured in the United States and Japan only and have been used not only in these countries but also in Europe and elsewhere. Their development in Japan, a country that depends entirely on imported raw salts for its soda industry, was aimed primarily at industrial-scale concentration of salt from seawater for this raw material. With the development of univalent-ion permselective membranes, which have a lower permeability to divalent ions present in seawater and perform effective concentration of sodium chloride, the seawater concentration process is able to operate on an industrial scale. Today, these membrane processes are in use to meet the domestic demand for table salt and are on the way toward practical applications for the supply of industrial salt.(2)

TABLE 1. TYPICAL APPLICATIONS OF ION-EXCHANGE MEMBRANES

Purpose	Method	Scale of Use
Production of table salt from seawater	Electrodialysis (concentration)	Commercial
Production of potable water from saline water	Electrodialysis (desalination)	Commercial
Treatment of radioactive waste	Electrodialysis (demineralization, concentration)	Pilot plant
Refining of blood serum and vaccine	Electrodialysis (demineralization)	Commercial
Treatment of dairy products	Electrodialysis (demineralization)	Commercial
Purification of amino acid	Electrodialysis (demineralization)	Commercial
Deacidification of fruit juice	Electrodialysis (deacidification)	Commercial
Treatment of sugar juice	Electrodialysis (demineralization)	Pilot plant
Purification of organic matter	Electrodialysis (demineralization)	Commercial
Recovery of waste acid	Dialysis	Commercial
Refining processes of nickel and copper	Dialysis	Commercial

Use of membranes that show permselectivity to ions having the same charge also enables separation of ions by their differences in valence and molecular weight, which has resulted in the more extensive application of electrodialysis. In the application of electrodialysis, studies not only of separation processes (such as concentration and desalination) but also of processes in which chemical reaction between ions permeating the membranes is effected have been carried out. This may be called an electrodialysis process with accompanying chemical reaction.

ELECTRODIALYSIS WITH ACCOMPANYING CHEMICAL REACTION

The principle of this process is illustrated in Figure 1. While conventional electrodialysis is a simple separation process, this new concept in electrodialysis is such that double-decomposition or substitution reactions occur between ions permeating ion-exchange membranes:

$$MX + NY \longrightarrow NX + MY \quad , \qquad (1)$$

where

M, N are cations
X, Y are anions.

Typical examples of chemical reactions studied are as follows[3]:

$$Ca(OH)_2 + 2NaCl \longrightarrow 2NaOH + CaCl_2 \qquad (2a)$$

$$NaNO_3 + KCl \longrightarrow KNO_3 + NaCl \qquad (2b)$$

$$Na_4Fe(CN)_6 + 4KCl \longrightarrow K_4Fe(CN)_6 + 4NaCl \qquad (2c)$$

While some of the anticipated applications of this process have already been introduced[4], very few are on a large scale as yet. One might cite as possible reasons for the delay in industrialization of the process, problems yet to be solved on permselectivity of ion-exchange membranes, process engineering, and operation cost. But, at the same time, the process offers the following possible advantages:

(1) Ability to perform reactions that cannot be achieved in ordinary reactors.

(2) Reaction products difficult to separate in ordinary reactors can be individually obtained and, also, the reactants and products can be handled separately.

(3) In case the reactants are dilute solutions, higher concentration of the products can be achieved by permeation through the ion-exchange membranes.

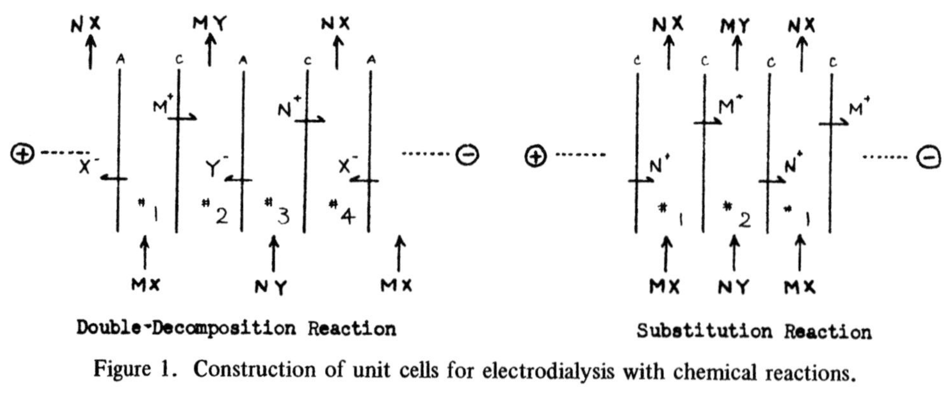

Figure 1. Construction of unit cells for electrodialysis with chemical reactions.

(4) By the use of membranes permselective to ions of the same charge, high-purity products can be obtained using raw materials in which impurities coexist.

As an example of a chemical reaction that can be carried out by this process, the production of acidic salt with a substitution reaction has been studied.

SUBSTITUTION REACTION FOR PHOSPHATE

Orthophosphates and pyrophosphates are utilized in various fields in the forms of alkali metallic and ammonium salts. Particularly, many acidic salts find extensive use in fields other than the chemical industry as chemicals for detergent and plating use or as food additives.

In the production of these salts, normal salts or acidic salts can be obtained by the neutralization of pure phosphoric acid with caustic alkali and ammonia. Also, acidic salts can be produced by the addition of pure phosphoric acid to pure salts. In either case, however, purified phosphoric acid and salts are essential. While it depends on the quality of phosphorite and the production process for phosphoric acid, purification is very difficult when phosphoric acid is produced not by the dry but by the wet process. It is almost impossible to obtain a pure acid, because a method of separating coexistent metallic ions, such as iron and aluminium or silica and fluorine, from phosphoric acid has not been developed. On the other hand, various techniques for removing these impurities in the production of phosphate have been studied, some of which have been industrialized.

Under the circumstances, we attempted to evaluate the substitution reaction by means of electrodialysis as a possible approach to the production of acidic salts without the use of pure phosphoric acid. Sodium orthophosphate was chosen as a component of the reaction system because it was felt that this process could be applied to the production of sodium tripolyphosphate. That is, sodium tripolyphosphate is produced by the calcination of acidic sodium phosphate, which in turn could be produced by the electrodialysis process.

The reactions performed by electrodialysis are as follows:

$$Na_3PO_4 + H^+ \longrightarrow Na_2HPO_4 + Na^+ \tag{3a}$$

$$Na_2HPO_4 + H^+ \longrightarrow NaH_2PO_4 + Na^+ \tag{3b}$$

As sources of hydrogen ion for these reactions, various acids including sulfuric acid and phosphoric acid are available.

PRELIMINARY EXPERIMENT

Characteristics of Phosphate

Figure 2 gives a titration curve showing the change in pH with the addition of caustic soda to pure phosphoric acid. The progress of the substitution reaction by means of electrodialysis can thus be checked according to changing pH value of phosphate.

As to solubility of phosphate, the information illustrated in Figure 3 is available.(5) The reaction is performed, desirably, at higher temperature and with reactant concentrations as high as possible, if the final product is to be obtained in the solid form.

The density of co-ion in the ion-exchange membrane tends to increase with increasing concentration of the outer solution, which reduces the membrane's permselectivity. Prior to operation, therefore, it is necessary to investigate current efficiency, cross leakage, etc., at high solution concentrations.

Possible advantages accompanying the high-concentration operation include the rise of limiting current density and the reduction of electrical resistance of the cell. We decided to carry out this experiment at about 50 C in consideration of the threshold solubilities of phosphates.

Experimental Equipment

Ion-Exchange Membrane. As the standard ion-exchange membrane for this study, Selemion CSV, which is produced by Asahi Glass Company on a commercial scale, was selected, together with cation-exchange membranes a little different from it in performance for comparison. The Selemion CSV is a strongly acidic cation-exchange membrane having univalent cation permselectivity; it uses PVC cloth as reinforcement.

Typical specifications of the Selemion CSV membrane are as follows:

Thickness, mm	– 0.26 to 0.31
Bursting strength, kg/cm^2	– 6 to 8
Areal resistance, $\Omega \cdot cm^2$	– 8.0 to 12.0
Transport number	– $t_{Na} > 0.92$ $t_{Ca^{2+} + Mg^{2+}} > 0.14$

Electrical resistance was measured with 1000-cps alternating current in a 0.5 N NaCl solution. The transport number of Na^+ was obtained from the membrane potential between 0.5 N NaCl and 1.0 N NaCl and that of $Ca^{2+} + Mg^{2+}$ was obtained from the electrodialysis of seawater at a current density of 2 A/dm^2.

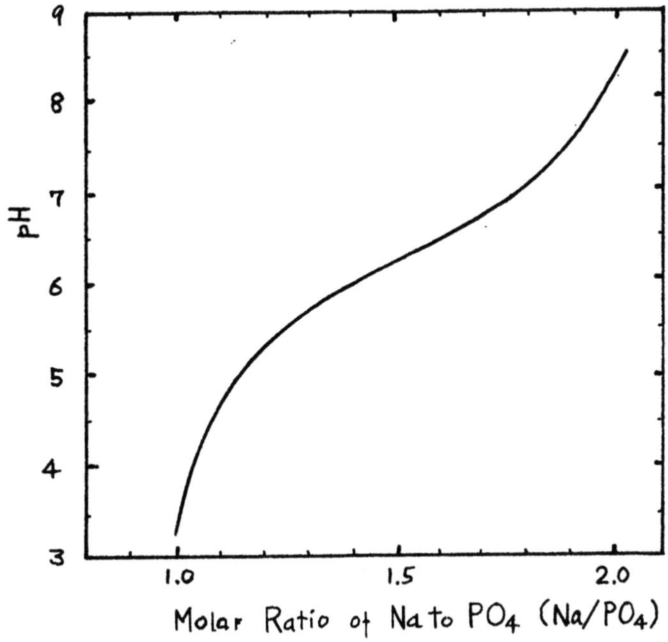

Figure 2. Titration curve of orthophosphoric acid.

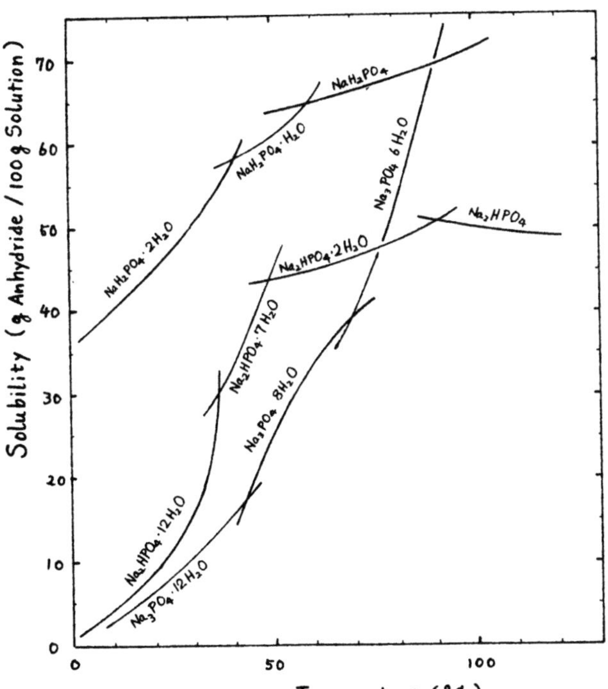

Figure 3. Solubilities of sodium orthophosphates.

Electrodialyzer. Designed so as to permit future scale-up to the industrial application level, the small-size experimental electrodialyzer used in this study has an effective membrane area of 209 cm^2 per unit cell, with membranes 2 mm apart. Each stack is composed of about 10 pairs of unit cells and is constructed as a filter-press type. Its main specifications are shown in Table 2.

The cell frame is made of chloroprene rubber, and polypropylene flow distributors are provided at both the solution inlet and outlet. In the cell is provided a polyethylene net-like structure to maintain membrane distance uniformly and to make the flow-velocity distribution uniform. The flow of the solution is a parallel upward flow.

Auxiliary Equipment. Auxiliary equipment was designed to enable both batch and continuous operation. Figure 4 illustrates the main piping system. To perform the experimental operation at about 50 C, the electrodialyzer is located in a thermostatically controlled air oven and each tank is heated by means of an immersion (electric) heater. The piping is composed of double pipes and kept warm with externally running hot water at about 50 C.

To enable a continuous check of the solution composition, its pH value is measured with a glass electrode placed in the tank. Also, the feed solution is fed by means of a constant-flow-rate pump in the continuous operation. Direct current is controlled by a silicon control rectifier provided with constant current or by a constant-voltage control device.

Experiment by Batch Method

To observe the progress of the substitution reaction with the acid salt, a series of experiments was carried out by the batch method using reagent-grade phosphoric acid and sulfuric acid. The reaction was performed at a constant current by the circulation of acidic sodium phosphate (Na_2HPO_4) and with the acid placed in the tanks in almost equal quantities.

TABLE 2. MAIN SPECIFICATIONS OF ELECTRODIALYSIS STACK

Type	DU-O_b
Size of Membrane, cm	24 x 16
Effective Area of Membrane Pair, dm^2	2.09
Thickness of Cell Frame, mm	2
Number of Membrane Pairs	10
Construction Material	
Cell Frame	CP Rubber
Flow Distribution	Polypropylene
Spacer	Polyethylene
Cathode	Stainless steel
Anode	Platinum-plated titanium

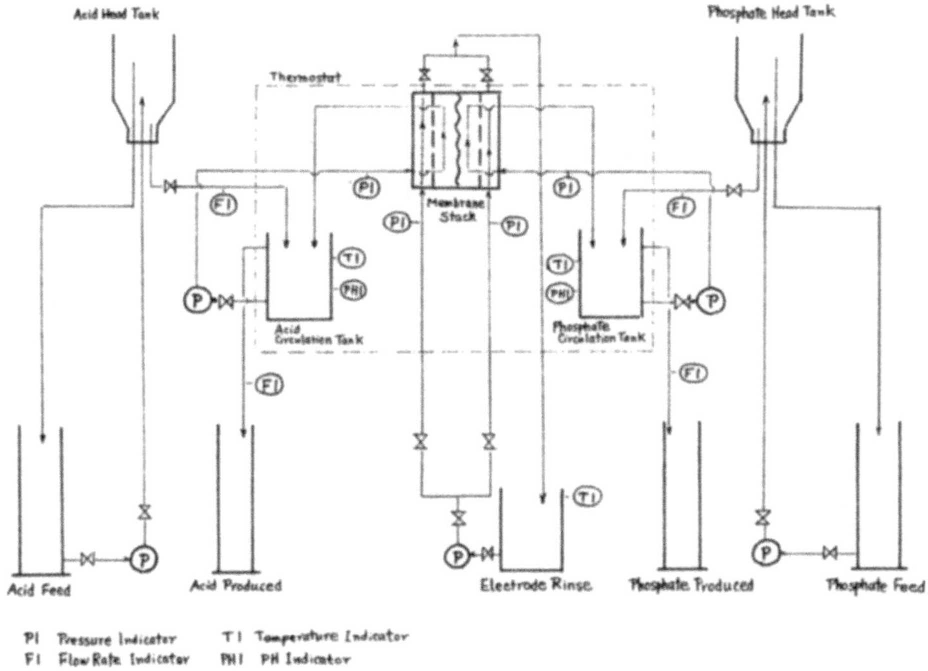

Figure 4. Experimental electrodialysis unit for the substitution reaction of sodium phosphate.

Depletion in pH Value of Sodium Phosphate. As mentioned above, Na_2HPO_4 was used as sodium phosphate for the substitution reaction with phosphoric acid under typical conditions as shown in Table 3. With the elapse of reaction time, the pH value of sodium phosphate took a downward trend such as shown in Figure 5. A comparison of this trend with the titration curve referred to earlier indicates the progress of the reaction.

Current Efficiency. As Table 3 indicates, current efficiency varied greatly in the different experiments. The current efficiencies were calculated relative to the substitution between Na^+ ion and H^+ ion in the pH-value range of sodium phosphate from 8.6 to 4.0, thereby relating them to the current densities as shown in Figure 6. As is noted, the maximum current-efficiency value is obtainable at current densities of 6 or 7 A/dm^2.

Electrodialysis in general gives rise to changes in pH value, simultaneous with the drop in current efficiency when the current density exceeds the limiting condition for the concentration polarization. In the case of an acid-base reaction, such as in this study, however, such changes in pH value would not occur and, thus, the current density associated with the maximum current-efficiency value, as mentioned above, may be regarded as the limiting current density.

TABLE 3. PERFORMANCE OF BATCHWISE REACTION

No.	Cell Voltage, V	Current Density, A/dm^2	Concentration: Sodium Phosphate (Na_2HPO_4), %	Concentration: Phosphoric Acid, (P_2O_5), %	pH Value of Sodium Phosphate: Initial	pH Value of Sodium Phosphate: Final	Current Efficiency, %
1	10.3	1.5	10	12.5	8.6	4.2	29.8
2	10.1	1.5	20	25	8.6	6.4	30.8
3	15.3	2.1	10	12.5	8.6	4.8	55.0
4	20.6	3.5	10	12.5	8.6	4.2	48.6
5	6.4	0.9	10	12.5	8.7	4.6	51.5
6	11.5	3.4	20	25	8.6	4.1	50.0
7	20.4	5.1	20	25	8.8	4.3	57.8
8	20.0	5.7	20	25	8.6	3.7	64.6
9	29.8	9.8	20	25	8.6	4.0	58.5
10	24.8	5.1	20	25	8.6	3.5	62.6

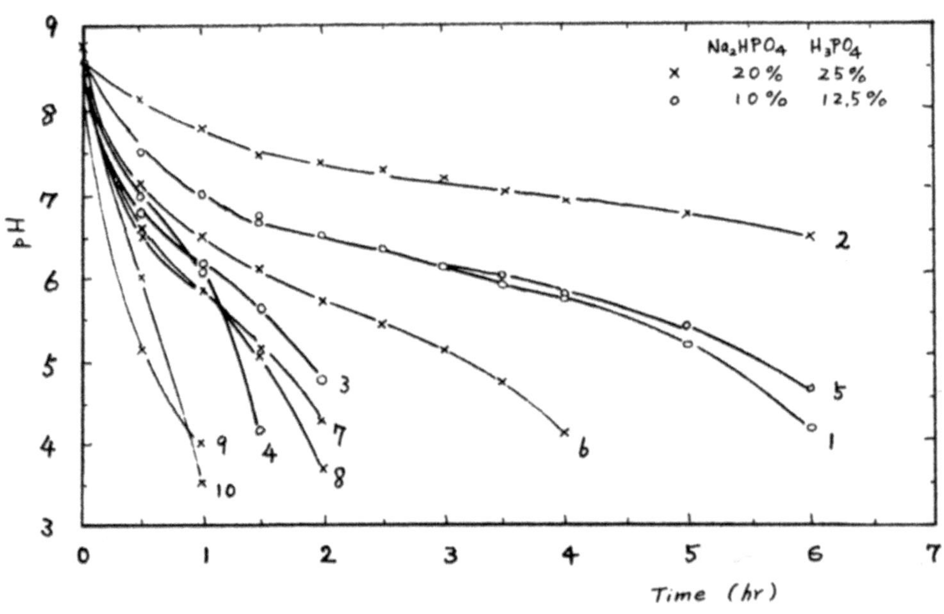

Figure 5. Depletion of pH of sodium phosphate in batch experiment (cf Table 3).

Acid Composition vs Current Efficiency. In the substitution reaction of sodium phosphate with phosphoric acid, the anions are the same, so their behavior is impossible to analyze. For the acid-composition experiment therefore sulfuric acid was used instead. In this experiment, 40 percent Na_2HPO_4 was continuously fed as sodium phosphate, with the solution flowing from the tank maintained at a pH of 6.4 to 6.6 (equivalent to $2Na_2HPO_4 \cdot NaH_2PO_4$). As for sulfuric acid, 40 percent H_2SO_4 was placed in the tank beforehand, and the reaction was carried out at 6 A/dm^2 by the batch method.

The composition of sulfuric acid used and the current-efficiency data obtained in this experiment are shown in Figure 7. When the composition of sodium phosphate was maintained constant, the current efficiency in the substitution reaction depended greatly upon the composition of the acid. When sulfuric acid was used, the findings indicated a remarkable change in current efficiency was likely to be associated with a sodium content approximately equal to that in $NaHSO_4$. Because of this, further study is deemed necessary in view of the rate of consumption of acid.

Experiment by Continuous Method

The batchwise experiment showed that the acidic salt can be obtained as anticipated, with a drop in pH value of sodium phosphate similar to the pattern of the titration curve. It was discovered that the acid consumption, current density, etc., were responsible for the drop in pH.

The substitution reaction between 40 percent Na_2HPO_4 and 40 percent sulfuric acid was carried out by the continuous method using the small-size electrodialyzer. Prior to this operation, it was planned that the acidic sodium phosphate to be produced would be so controlled, by maintaining the pH at 6.6 to 6.4, that its composition would permit the production of sodium tripolyphosphate by calcination, and that 40 percent of the sulfuric acid would be consumed. The progress of the reaction conducted at 50 C at a current density of 3 A/dm^2 is shown in Figure 8.

Analysis of the typical product overflow from the circulation tank gave data shown in Table 4, which indicate a certain leakage of anion. The current efficiencies observed for the substitution reaction were

$$94 \text{ percent relative to } H^+ \text{ ion}$$

$$92 \text{ percent relative to } Na^+ \text{ ion} \quad ,$$

showing a slight difference. The transport numbers of the ions, obtained on the basis of their material balance and transmission through the ion-exchange membranes, are given in Figure 9. In this experiment, it was noted that loss of phosphate ion was small, but mixing of sulfate ion into phosphate occurred to a certain extent. Should these impurities give rise to problems because of product specifications, further study would be called for. In many cases, however, the leakages are not to such an extent as to affect the quality of the product. As indicated above, the current efficiency in the substitution reaction exceeded 90 percent, which is very high.

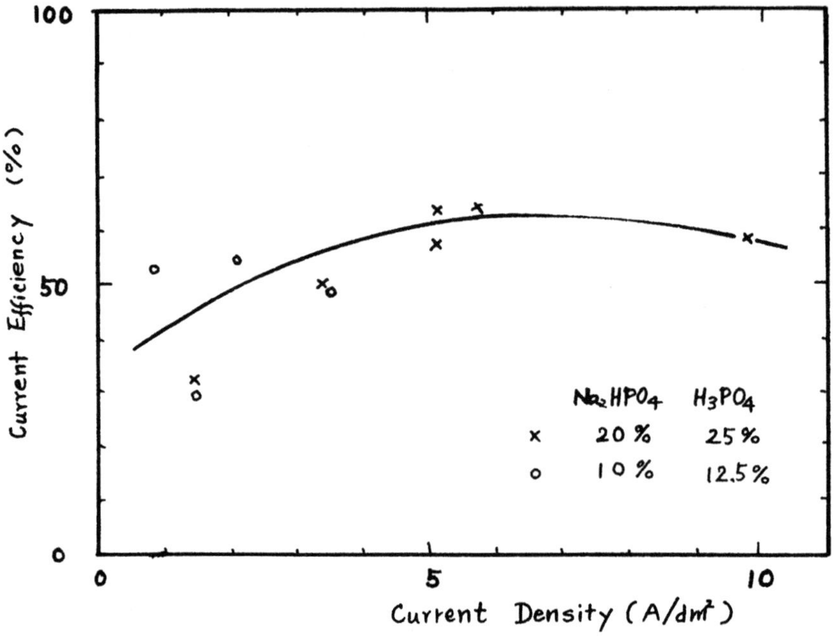

Figure 6. Relation between current density and current efficiency in batch experiment.

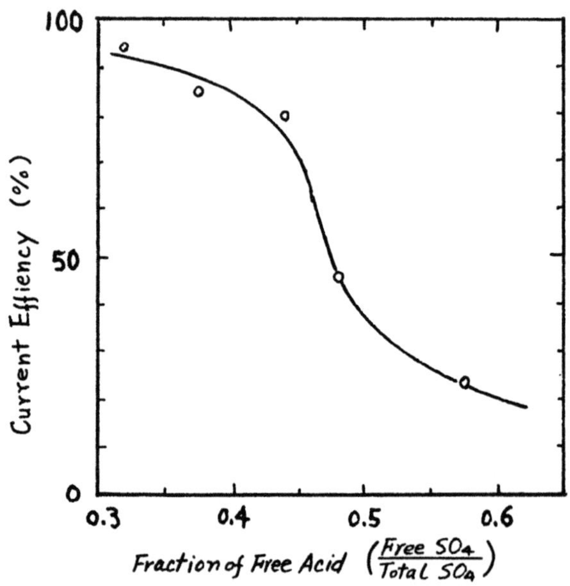

Figure 7. Relation between current efficiency and the composition of acid solution.

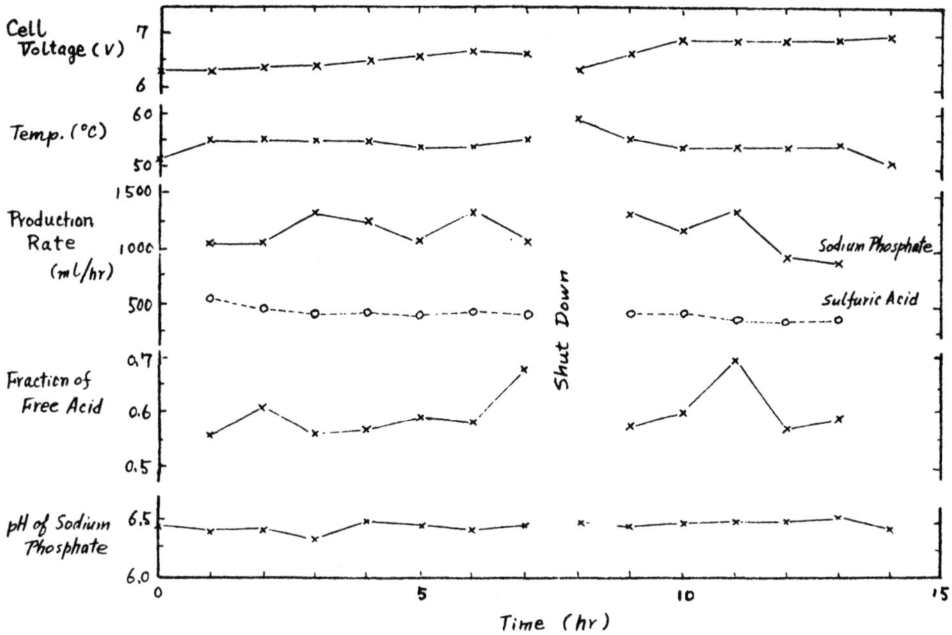

Figure 8. Result of continuous reaction of sodium phosphate and sulfuric acid (3 A/dm²).

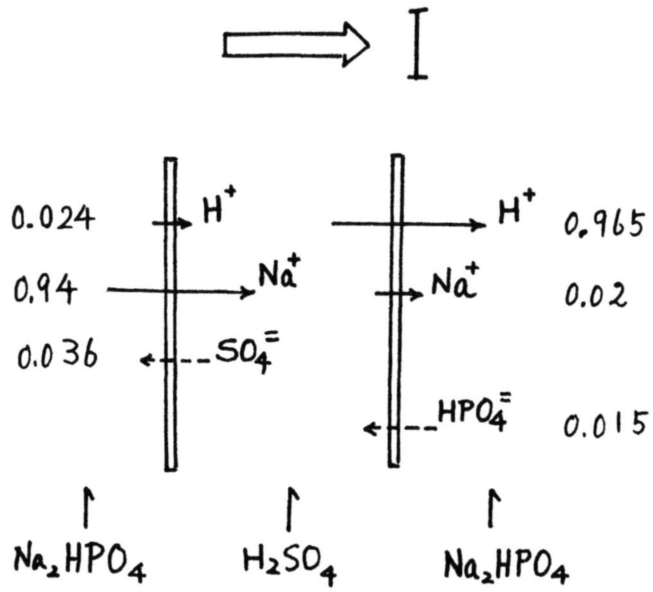

Figure 9. Transport numbers of the ions in the electrodialysis of sodium phosphate and sulfuric acid.

TABLE 4. TYPICAL COMPOSITION OF SUBSTITUTION REACTION PRODUCTS (CONTINUOUS OPERATION, 3 A/dm^2)

Component	Sodium Phosphate 1	Sodium Phosphate 2	Phosphoric Acid 1	Phosphoric Acid 2
P_2O_5, %	22.2	21.9	0.14	0.17
Na, %	12.3	12.2	7.0	6.8
SO_4, %	0.16	0.16	34.1	33.6
Specific gravity	1.42	1.42	1.34	1.34

Substitution Reaction With Wet-Process Phosphoric Acid

Successful industrialization of the substitution reaction by means of electrodialysis requires resolution of the problem as to the choice of suitable acid as the source of hydrogen ion. Such an acid must be obtained commercially and be of reliable quality, and the anion constituting that acid should permeate the membranes with difficulty, or slight leakage of that anion should not affect the quality of acidic sodium phosphate produced. Furthermore, it is desirable from the economic point of view that the acid used should provide a sodium salt of useful and valuable quality. As a result of careful study of availability and other possible requirements, we decided to use wet-process phosphoric acid for this experiment.

Wet-process phosphoric acid is purified by the addition of caustic soda and is finally used for the production of sodium phosphate. Its use thus offers the advantage, as shown in Figure 10, that alkali consumption can be saved by adding some of the required caustic alkali to it during the substitution reaction prior to the production of sodium phosphate.

Experimental Method

Selemion CSV membranes were installed in the small-size electrodialyzer used in previous experiments, and the substitution reaction was performed continuously with constant current by means of the equipment shown in Figure 4.

We employed an experimental method identical with the continuous method used in the preliminary experiment.

Wet-Process Phosphoric Acid

The typical composition of phosphoric acid used in this experiment is shown in Table 5. Addition of Na^+ ion to it precipitates sodium silicofluoride, at a pH near 0.5, and gives iron and aluminium in the form of phosphate at a pH of about 1.8.

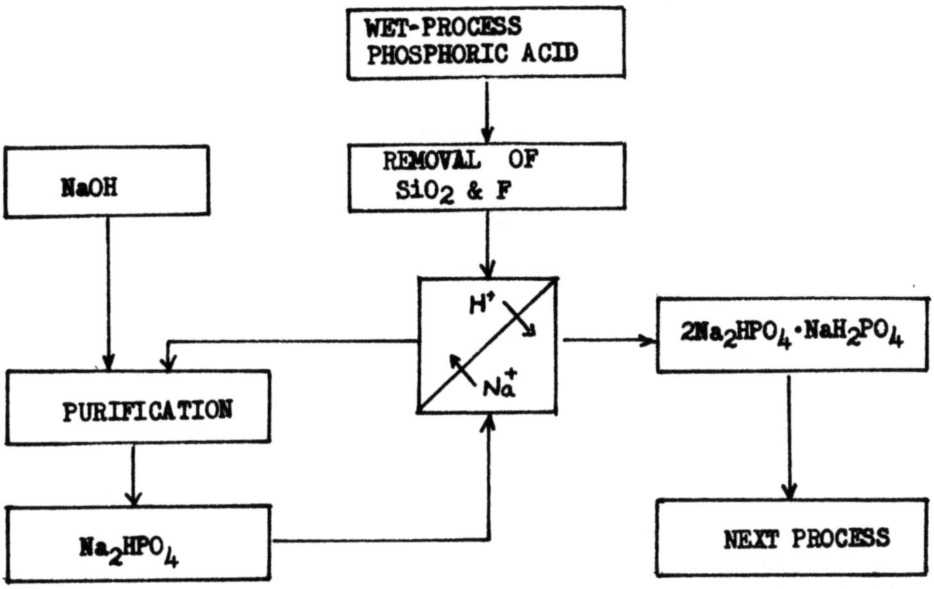

Figure 10. Flow diagram of substitution reaction of sodium phosphate and wet-process phosphoric acid by means of electrodialysis.

TABLE 5. TYPICAL COMPOSITION OF ORTHOPHOSPHORIC ACID (PRODUCED BY WET METHOD)

Constituent	Percent
P_2O_5	27-28
SO_4	4-5
SiO_2	1.3
F	2.5
Fe	0.5
Al	0.6
Ca	0.2
Mg	0.3

In order to prevent scale formation in the electrodialyzer, therefore, the substitution reaction should be carried out using phosphoric acid from which the sodium silicofluoride has been removed by raising its pH value to about 0.5, and by maintaining the pH after reaction below 1.8.

Experiment

While maintaining the pH's of the acidic sodium phosphate and phosphoric acid obtained from this reaction at 6.6 and 1.8, respectively, experiments were conducted with current densities at 2 and 3 A/dm^2. Typical results of the experiment at 3 A/dm^2 are shown in Figure 11, and the experimental conditions responsible for such data are as follows:

Feed concentration of sodium phosphate (Na_2HPO_4), percent	40
Feed concentration of phosphoric acid (P_2O_5), percent	29
Current density	3 A/dm^2

This particular experiment was run continuously for 16 hours, during which time no solid formation was detected in the electrodialyzer or in the solution. Reaction in this experiment was effected with

$$3Na_2HPO_4 + H_3PO_4 \rightarrow 2Na_2HPO_4 \cdot NaH_2PO_4 + NaH_2PO_4 \quad , \qquad (4)$$

where current efficiency was maintained above 93 percent. The power consumption was registered as about 87 kwh/t at 2 A/dm^2 and about 112 kwh/t at 3 A/dm^2.

The compositions of the samples used and products obtained in this experiment are shown in Table 6.

Scale Formation

On the strength of the successful, but relatively short-time continuous operation, we repeated the experiment over a prolonged period under varied conditions of current density, flow velocity, etc.; it was found that a white-colored scale formed at times on the sodium phosphate side of the membranes through which hydrogen ion permeated. Although this caused no abnormalities during the operation, the reason for its formation was difficult to explain.

Analysis of the scale formed, given in Table 7, indicated it to be NaH_2PO_4, which was confirmed by X-ray crystal analysis. From the fact that the pH and the concentration of the solution did not favor the formation of NaH_2PO_4, it is presumed that the scale formation resulted from accelerated crystallization due to crystal nucleus and/or other reasons.

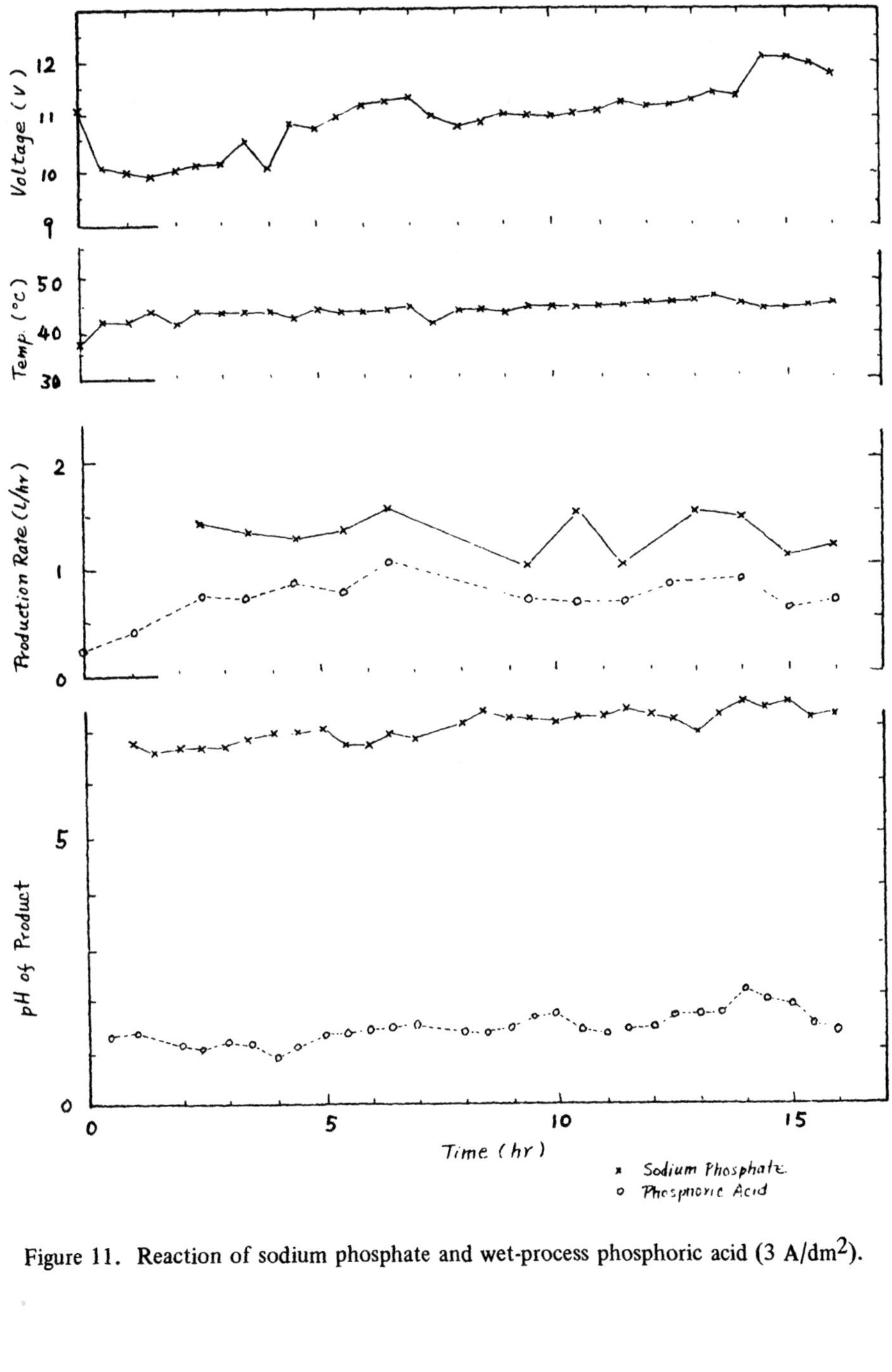

Figure 11. Reaction of sodium phosphate and wet-process phosphoric acid (3 A/dm^2).

TABLE 6. TYPICAL COMPOSITIONS OF REACTANTS AND PRODUCTS (CONTINUOUS OPERATION, 3 A/dm^2)

Constituent	Phosphoric Acid		Sodium Phosphate	
	Reactant, %	Product, %	Reactant, %	Product, %
P_2O_5	28.1	25.80	21.0	21.4
Na	1.91	3.55	13.7	12.3
SiO_2	0.044	0.055	0.034	0.026
F	0.46	0.33	0.0008	0.0057
Fe	0.43	0.39	0.0004	0.0012
Al	1.63	1.38	0.0043	0.0049
Ca	0.152	0.084	Trace	0.0005
Mg	0.106	0.114	Trace	0.0026
SO_4	1.80	1.62	Trace	0.018

TABLE 7. COMPOSITION OF SCALE FORMED ON SELEMION CSV IN SODIUM PHOSPHATE CHAMBERS

Reaction between sodium phosphate and wet-method phosphoric acid, 3 A/dm^2.

Constituent	Percent
P_2O_5	48.5
Na	13.2
SiO_2	0.3
F	0.6
Fe	0.2
Al	0.9
Ca	0.4
Mg	0.2

Experiment on Scale Formation. In the middle of the four-chamber type of cell with an effective membrane area of 64 cm were inserted the Selemion CMV membranes as ordinary cation-exchange membranes bearing no univalent cation permselectivity. In both side chambers were placed Na_2HPO_4 and phosphoric acid, and electricity was fed to the pair of electrodes set in both end chambers of the cell at a current density of 3 A/dm^2 at a temperature of 55 C.

Investigations were carried out for the presence or absence of scale formed by the uses of different kinds of phosphoric acid, with the addition of various types of cations to reagent-grade phosphoric acid in quantities equal to their content in wet-process phosphoric acid. Results were as follows:

Test	Phosphoric Acid	Cation Added	Scale Formation
1	Wet-process phosphoric acid	None	+
2	Reagent-grade phosphoric acid	None	–
3	Ditto	Na^+	–
4	"	Na^+, Ca^{2+}	+
5	"	Na^+, Al^{3+}	–
6	"	Na^+, Fe^{2+}	–
7	"	Na^+, Mg^{2+}	–

It was found that scale formation occurred only when Ca^{2+} ion was added. Analysis of scale formed in Tests 1 and 5 is shown in Table 8, which indicate its correspondence to NaH_2PO_4. From this, it is presumed that the scale formation is caused by the slight amount of permeation of Ca^{2+} through H^+-ion-permselective membranes.

Permselectivity With Univalent Cation. Led by the above experiments to the conviction that scale formation could be prevented by controlling the permeation of Ca^{2+} ion through the H^+-ion-transmissible membranes, we proceeded to study the Ca^{2+}-ion permselectivity of the Selemion CSV membranes and ion-exchange membranes similar to CSV manufactured by changing the cross-linkage of the resin only.

Conventionally, the univalent-cation permselective membrane is evaluated in respect to the permeability ratio between Na^+ ion and divalent ions in a neutral solution. However, since this experiment involved the permeability ratio between H^+ ion and Ca^{2+} ion, comparative investigations were carried out on the transport number of Ca^{2+} ion and scale formation, treating the two kinds of solutions below:

(1) 0.4 N HCl + 0.1 N $CaCl_2$

(2) 0.4 N NaCl + 0.1 N $CaCl_2$.

TABLE 8. TYPICAL COMPOSITION OF SCALE FORMED IN SODIUM PHOSPHATE CHAMBERS[(a)]

Phosphoric Acid	Composition, percent Wet Method, Test 1	Dry Method (Na^+ and Ca^{++} Added), Test 5
P_2O_5	47.9	52.8
Na	15.0	8.1
SiO_2		0.062
F	0.136	0.18
Fe	0.174	0.029
Al	0.24	0.02
Ca	0.27	1.22
Mg	0.33	0.00
SO_4	0.60	0.07

(a) Membrane: Selemion CMV
Cell: Four-chamber-type cell.

The transport numbers of the ions in these solutions were measured at 25 C at a current density of 2 A/dm^2. Characteristics of the membranes used in the experiments are shown in Table 9. These data indicate that, while the transport number of Ca^{2+} ion of the HCl-$CaCl_2$ system showed a positive correlation with scale formation, that of Ca^{2+} ion in the neutral solution did not do so consistently. This means that the Selemion CSV membranes should be such that the transport number of Ca^{2+} ion is below 0.01 in the above acidic solutions when they are applied to practical use. Particularly, some membranes, such as those in Test 6, do not show any remarkable permselectivity in a neutral solution but excellent performance in an acidic solution, and it goes without saying that membranes equal to those in performance are desirable.

Confirmation by Continuous Experiment. After finding that scale formation could be prevented by the improvement of the permselectivity of the ion-exchange membranes to ions of the same charge, a continuous experiment was carried out for about 40 hours with the small-size experimental electrodialyzer in order to confirm these findings.

The experimental conditions were the same as conventional in that 40 percent Na_2HPO_4 and wet-process phosphoric acid containing 26 percent P_2O_5 were used as raw solutions and the operating temperature and current density were 55 C and 3 A/dm^2, respectively. The pH's of sodium phosphate and phosphoric acid were maintained at 6.6 and 1.5, respectively.

The ion-exchange membranes were Selemion CSV membranes equal in performance to those evaluated in Test 6. Together with them, two sheets of membranes from the lot that

TABLE 9. CHARACTERISTICS OF UNIVALENT-CATION PERMSELECTIVE MEMBRANES (SELEMION CSV)

Lot	Areal Resistance, Ω-cm^2	Thickness, mm	Transport Number, Ca^{2+}: 0.4 N HCl +0.1 N $CaCl_2$	Transport Number, Ca^{2+}: 0.4 N NaCl +0.1 N $CaCl_2$	Scale Formation
1	7.7	0.303	0.015	0.087	–
2	7.4	0.294	0.012	0.133	+
3	7.7	0.293	0.058	0.127	+++
4	9.9	0.264	0.051	0.252	++
5	6.8	0.303	0.010	0.017	±
6	12.1	0.225	0.0007	0.024	–

allowed a trace of scale to be formed in a previous series of experiments were also employed for comparison purposes.

The experimental operation lasted for about 40 hours, and the development during that period was as shown in Figure 12. There was observed no unusual phenomenon, except a voltage variation of a certain degree in the course of operation. Upon completion of the experiment, the experimental electrodialyzer was dismantled. No scale had been formed on the membranes corresponding to Lot 6, but some had formed on the two sheets of the comparative CSV, supporting our findings above.

CONCLUSIONS

It was found that, by the use of the small-size electrodialyzer, acidic sodium phosphate suitable for calcining to sodium tripolyphosphate could be produced through a substitution reaction between sodium phosphate (from purified wet-process phosphoric acid) and raw wet-process phosphoric acid. Based on these data, a series of pilot-scale tests were carried out by means of an electrodialyzer in a medium-scale commercial plant.

As a result, design data for electrodialysis facilities were obtained, together with helpful information on their combination with preprocessing and postprocessing, thus clarifying problems faced in the industrialization of the process for acidic-salt production. But in this report, there is no reference to industrial development of this process beyond the pilot-plant stage.

While the substitution reaction by means of ion-exchange membrane electrodialysis has begun to find applications in the deacidification of fruit juices and some other chemical processes, there are many more difficult technological problems yet to be solved than for its application to desalination and concentration. These problems must be solved in individual cases, and what we can state at this moment is that they are generally characterized by the need to prevent cross leakage of raw solutions in the electrodialyzer and the solid formation of reaction products or by-products in the membrane stack. For that matter, this report may be

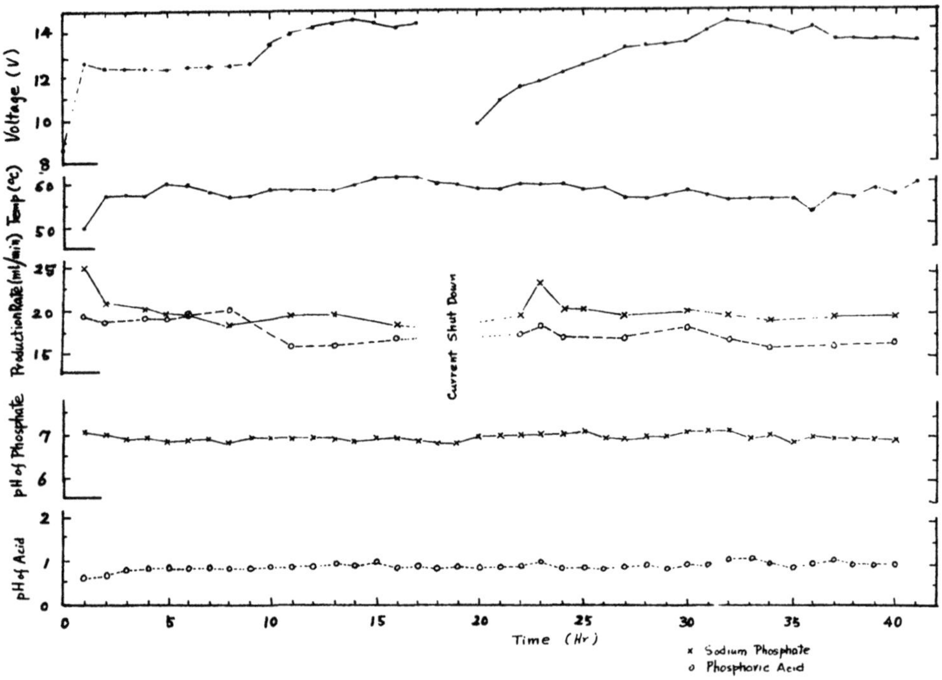

Figure 12. Reaction of sodium phosphate and wet-process phosphoric acid by means of highly permselective membrane (3 A/dm^2).

regarded as introducing one solution to such problems, and it is believed that further research-and-development efforts of this sort will certainly enhance the possibility of chemical reaction by means of electrodialysis and pave the way for its practical application to various fields in the near future.

REFERENCES

(1) T. Nishiwaki and S. Itoi, *Japan Chemical Quarterly*, **5** (1), 36-40 (1969).

(2) T. Nishiwaki and S. Itoi, ibid, **5** (4), 34-40 (1969).

(3) S. Uemura, T. Yawataya, and H. Toyabe, Japan Patent 233,003.

(4) E. A. Mason and W. Juda, *C.E.P. Symp. Ser.*, **55**, No. 24, 155-62 (1959).

(5) J. R. Van Wazer, *Phosphorus and Its Compounds*, Vol. 1, **500** (Interscience Publishers Inc., New York, 1958).

HYDROMETALLURGICAL SEPARATIONS BY SOLVENT MEMBRANES

Dr. René Bloch
Hydronautics-Israel, Ltd.
Rehovot, Israel

Comparing the schedule of the present symposium on industrial membrane processes with similar meetings a few years ago, i.e., the NATO Symposium on Membrane Applications in Ravello in 1966, one is positively surprised at the progress that has been made and also to what extent was justified the optimism of those who foresaw a few years ago a new chemical unit operation in the field of membrane separations. Reverse osmosis is today a leading process in the desalination of brackish water. Its use in various other industrial applications has successfully been proven in various installations of pilot-plant size. The same seems to be true of various ultrafiltration processes in the pharmaceutical industry and in food processing. The expectations attached today to the membrane field are well demonstrated by the exclusive forecast of the Mellon Institute which foresees by 1975 a five-fold increase of the annual expenditure volume in membrane rates, presently estimated at 15 million dollars. When mentioning this list of successful membrane processes one should, as a matter of fairness, mention some of the processes which have not completely lived up to the expectations held a few years ago. It is probably no coincidence, but representative for the present state of the art, that processes such as liquid permeation and gas separation, both of which were once considered of large potential interest to the oil-processing industry, do not appear on the program of the present symposium.

Our group in Rehovot is active in a number of reverse-osmosis and ultrafiltration processes discussed extensively at this symposium. We have chosen, however, to review a category of membrane separations which has received relatively little attention outside of Israel, but which within Israel is still pursued with considerable effort – e.g., by three separate groups working in parallel, namely, the field of hydrometallurgical separations by solvent-type membranes. We cannot foresee at this stage whether this process will share the fate of liquid permeation or gas separations by membranes or whether it will prove its practical feasibility in the near future. Whatever its practical future, we hope to be able to show that solvent-type membranes for metal ions offer a fascinating subject for membrane studies which have led to the observation of unexpected membrane phenomena of rather wide scientific significance.

We shall start by defining the term "solvent membranes for metal ions" for the purpose of this presentation. We shall then give a description of a number of phenomena characteristic of such membrane processes. In particular, we shall demonstrate the presence of a rather new selectivity principle in membranes, namely, differences in the rate of association or dissociation of a permeating species with the reactive sites of the membrane matrix, a phenomenon that was forecast by M. Eigen a few years ago. We shall present experimental evidence for the validity of this concept and – deviating from the subject of this presentation – show other cases where the magnitude of permeability processes seems to be determined by reaction rates. We shall conclude by giving some speculative thoughts to the scientific and practical significance of solvent membranes with permeability that is reaction-rate controlled.

The large numbers of membranes investigated can be classified into three distinct groups:

(1) Membranes with permeability and selectivity determined by pore size and molecular size of the permeating species. This type includes ultrafiltration membranes (extensively discussed at this meeting), common cellophane-dialysis membranes, and microfilters with pores in the micron range.

(2) Membranes with matrices that are electrically charged and the permeabilities of which are determined by the electrostatic interactions of the permeating species with the charged matrix. Electrodialysis membranes, many biological membranes, and the hyperfiltration membranes of the Oak Ridge group belong to this category.

(3) Membranes that enable the permeation of a substance by chemically dissolving it, such as cellulose acetate membranes for reverse osmosis and membranes used in liquid permeation and separation of gases.

As previously mentioned, we are going to report in this lecture on the study and development of solvent membranes in a field that apparently has attracted relatively little attention until now: solvent-type membranes for metal complexes. The purpose of this study is twofold. First, the significance of the permeation of metal complexes through quasi-liquid membranes, i.e., the lipid layer of the cell wall, for the understanding of membrane phenomena in biological systems has been recognized since the studies by Overton and others.(1) Second, solvent extraction as a hydrometallurgical separation process has assumed a more and more dominant role in recent years. This is based on the fact that certain organic solvents, immiscible with water, show a high solubility and selectivity for certain metal salts and metal complexes. It was the purpose of the investigation carried out in Rehovot to study and develop membranes that reproduce the selective solubility of solvents used in hydrometallurgical separations.

The link between permeation through complexing solvent membranes and solvent extraction is easily seen in the following model experiment. In the U-tube shown in Figure 1A, the aqueous phases A and B are separated by a solvent plug S (heavier than water). If, among the various ions contained in solution A, there will be a certain species soluble in S, it will pass into the solvent phase and then reemerge into the aqueous phase, completing a solvent-extraction cycle.

Suppose now that the solvent plug is made thinner and thinner until its thickness is reduced to the dimensions of a membrane. The same experiment would then represent a membrane separation process (Figures 1B and 1C).

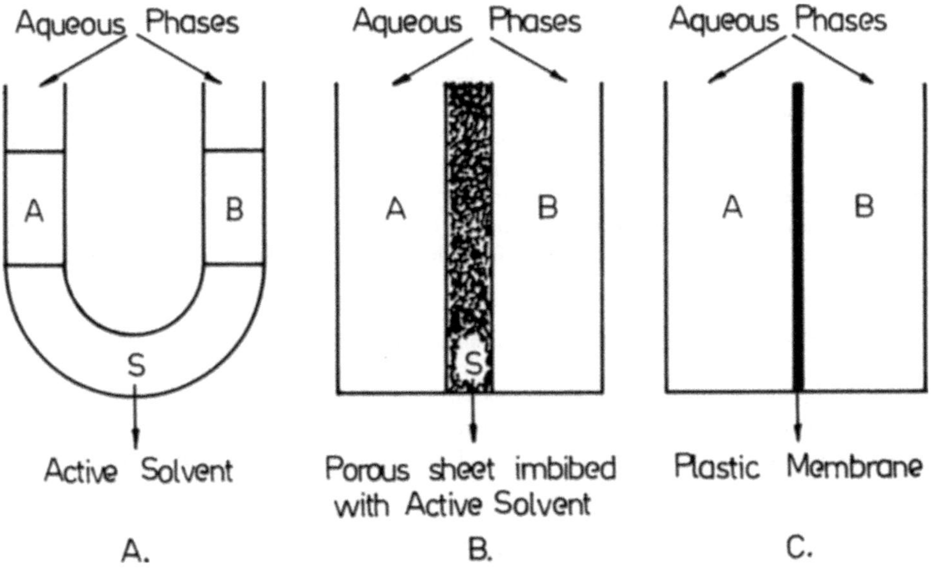

Figure 1. Link between permeation through complexing solvent membranes and solvent extraction.

It is not a new concept that the permeation mechanism of a number of membranes, in particular biological membranes, can be understood by studying dissolution and transport phenomena through a liquid phase. Overton, mentioned previously, demonstrated near the turn of the century that the permeability of various small molecules (in particular, narcotics) through lipid membranes is parallel to their solubility in oils or, more exactly, to the distribution coefficients of these narcotics between water and nonpolar solvents.

Overton, Schulman, Sollner, and Ilani[3-5] dealt essentially with systems that could be considered as models for membranes and thereby reached conclusions as to certain permeation phenomena through real membranes. It is, however, possible to go one step further and prepare plastic films which duplicate the solution behavior of certain solvents. This may be of particular interest if one succeeds in imitating solvents which are used for the extraction of precious metals. These hydrometallurgical processes are, as is well known, rapidly gaining importance among recovery processes for such metals.[2]

There are two basically different techniques used to construct membranes with matrices imitating such solvents: (a) a film-forming polymer having a monomeric unit which shows the greatest possible similarity to an extracting solvent may be synthesized[6]; (b) if the selective extractant has a high boiling point and low solubility in water, it may be used as a plasticizer for a suitable plastic.[7] The plastic material in this case only serves as a mechanical support for the immobilized selective solvent, forming a highly swollen organic gel.

The former possibility, i.e., the use of selective polymers, can be exemplified by a membrane consisting of polyethyl acrylate. This material reproduces the selective solubility of ethyl propionate for iron chloride with respect to aluminum chloride in concentrated hydrochloric acid. It forms a membrane selectively permeable to ferric tetrachloro acid ($HFeCl_4$).

The second technique is represented by a membrane consisting essentially of tributyl phosphate (TBP) and used for the recovery of uranyl nitrate from a model leach containing, in addition to nitric acid, ferric nitrate and aluminum nitrate. In this technique, use was made of the "happy coincidence" that TBP, the well-known extractant, is a common plasticizer for a number of plastics. Permeability rates comparable to the dialysis rates of common salts through cellophane have been measured through such membranes. Selectivity coefficients for uranyl nitrate with respect to nitric acid were rather low. With respect to sodium, iron, and aluminum nitrates, values were obtained which, to the best of our knowledge, are considerably larger separation factors than those previously reported for membranes distinguishing between ions of equal valency and equal charge. They reflect, however, selectivity coefficients obtained in solvent-extraction separation.

Let us point out some characteristic phenomena of solvent-type membranes:

Permeation through solvent membranes is known to be slow. For a simplified case, the permeability coefficient can be considered to be the product of the distribution coefficient of the permeant between membrane matrix and aqueous phase (K) and its diffusion coefficient in the membrane (D):

$$P = \frac{K \cdot D}{d} \tag{1}$$

where d denotes the thickness of the membrane. The case where $K > 1$ seems to be realized rarely, and D is usually considerably smaller than in liquids, as we are dealing at best with swollen gels if not with solid membranes. For this reason, appreciable permeation can only be observed in very thin membranes.

In order to obtain large transfers, ways have to be found to drastically increase the driving force above the value determined by the magnitude of the concentration difference in near ideal solutions. In hyperfiltration and in gas separations through membranes, this is achieved by applying high pressures.

We shall show in the following that, in the case of solvent membranes for metal complexes, an analogous increase of the driving force for permeation can be achieved by an impermeable additive that causes a drastic shift in the complexation equilibrium in favor of the complex and therefore greatly increases the activity difference of the permeant.

Since separation by means of a solvent membrane is closely related to solvent extraction, both processes will be characterized by common parameters, namely, a distribution coefficient (K) and a metal-complex association constant. This latter constant is important, as in most practical cases the species being extracted (or transported across the membrane) is a metal-ligand complex.

However, unlike solvent extractions, an additional parameter – a diffusion coefficient in the membrane, D – has to be introduced in order to describe membrane permeation. Consider first the case where the only species permeating the membrane is a stoichiometric metal complex (Figure 2). It is assumed that each of the membrane surfaces is in rapid equilibrium with its adjacent solution S^I and S^{II} and that transport across the membrane is entirely diffusion controlled. Hence, the concentrations of a permeant i at the two surfaces (just inside

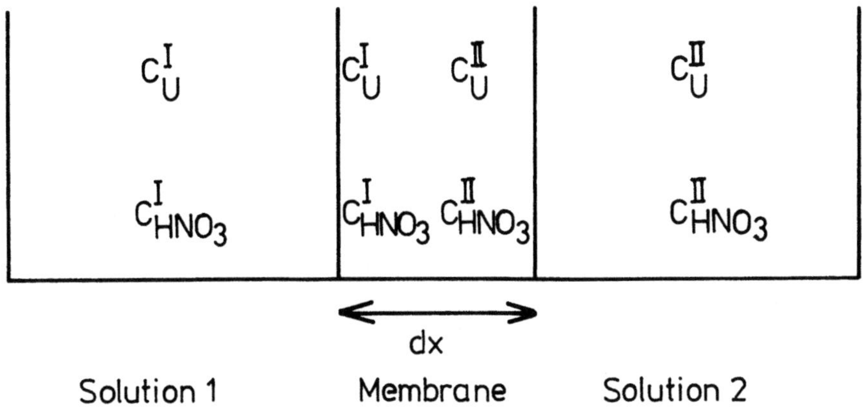

Figure 2. Uranium concentrations in membrane and contacting solutions.

the membrane) will be completely determined by two distribution coefficients K^I and K^{II} and the concentrations of the permeant in the respective solutions.

We assume a membrane of thickness d dividing two aqueous compartments containing a metal species i in the concentration c_i^I and c_i^{II}. The distribution coefficients between the aqueous phases and the membrane phase will be denoted $K_i^I = C_i^I / c_i^I$ and $K_i^{II} = C_i^{II} / c_i^{II}$ where capital C's denote metal-complex concentrations just inside the membrane at two surfaces.

The flux J_i from S^I into S^{II} is then given by

$$J_i = \frac{D}{d}\left(C_i^I - C_i^{II}\right) = \frac{D}{d}\left(K_i^I c_i^I - K_i^{II} c_i^{II}\right) \qquad (2)$$

It is well known that the distribution coefficient of metal association complexes between an aqueous and an organic phase may change by several orders of magnitude as a function of acidity or metal-ligand concentration. Two arbitrary illustrations of this phenomenon can be seen from Figure 3. Figure 3a represents data by Bankmann and Specker[8] who measured the distribution of $FeCl_3$ in different solvents as a function of the concentration of hydrochloric acid. It is seen that the distribution coefficient can range between less than 10 and 10^6. Another example showing less dramatic changes is shown in Figure 3b, with a system much closer to that of the membranes discussed here, namely, the extraction of uranyl nitrate into tributyl phosphate diluted with kerosene.[9]

It is possible to choose conditions in S^I and S^{II} such that K^I assumes a large value, while K^{II} is close to zero. As a consequence, the driving force for transfer through the membrane is considerably greater than reflected by the concentration difference of the permeating species between the two sides of the membrane. We referred earlier to this possibility of drastically increasing the difference in the chemical potential when we mentioned that a separation process based on diffusion is too slow unless ways (such as use of hydrostatic pressure or electric field) can be found to increase its driving force.

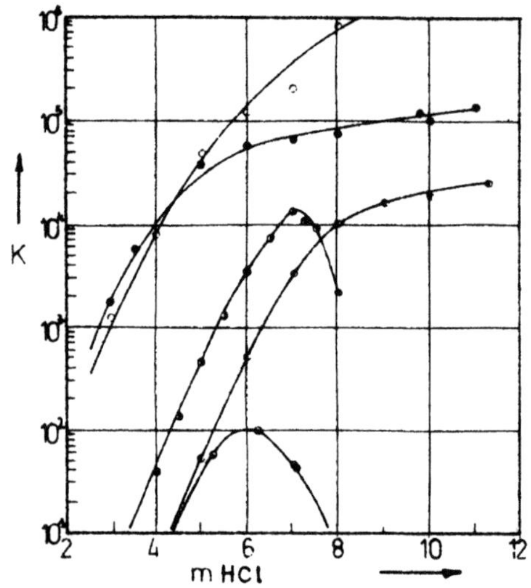

Figure 3a. Distribution of $FeCl_3$ in different solvents.[8]

◒ = Diethyl ether, ◑ = Isobutyl methyl ketone, ◐ = Di-n-propyl ketone, ● = Tributyl phosphate, O = Trianyl phosphate

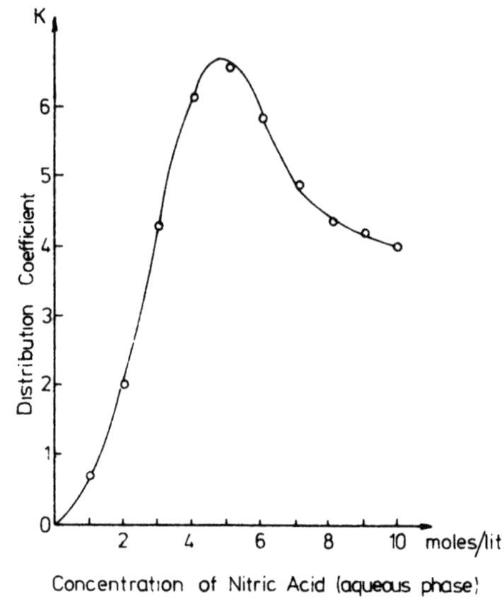

Figure 3b. Distribution coefficient of uranyl nitrate between nitric acid and tributyl phosphate diluted with hexane.[9]

A few results should demonstrate the nature and size of this phenomenon via the example of a membrane permeable to uranyl nitrate $UO_2(NO_3)_2$.

Figure 4 shows permeabilities for uranyl nitrate, i.e., $J_u/c_u^I = \frac{D_u \cdot K_u}{d}$ for a typical solvent membrane, e.g., of TBP as a function of concentration of nitric acid in S^I as well as corresponding values of K_u^I. A few words here about the experimental procedure:

> The membranes were prepared by coating with a coating knife a 10 percent solution of polyvinylchloride in cyclohexane mixed with three parts TBP per one part PVC on paper of high wet strength. After evaporation of the solvent, a coating of 40 microns' thickness remained. Permeation rates recorded represent initial values after reaching stationary state such that a change of the concentration difference between the two compartments in the course of the experiment did not have to be taken into account.

We see in Figure 4 a very close correlation between permeability rate and distribution coefficient. The permeability coefficient varies from practically zero to $150 \cdot 10^{-6}$(cm/sec) while the distribution coefficient reaches values as high as 85.

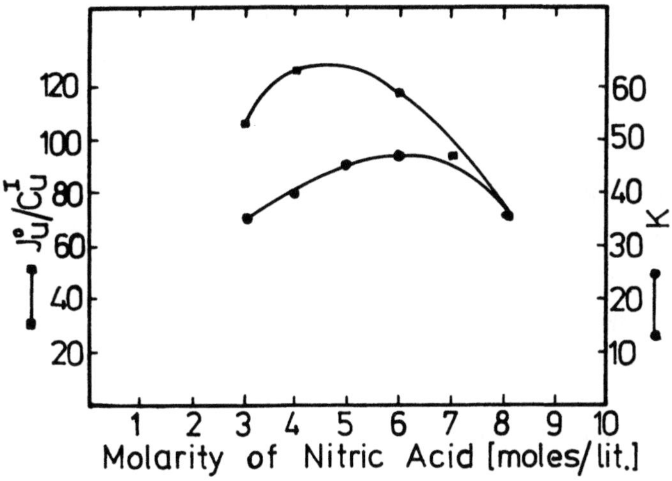

Figure 4. Uranium flux and distribution coefficients versus nitric acid concentration.(7)

Table 1 shows fluxes of uranyl nitrate in comparison with the flux of other metal species. We see that, apart from nitric acid, which shows an appreciable permeability, all other nitrates show extremely small permeability. The selectivity coefficients for sodium, ferric and aluminum ions assume values from 650 to 60,000, which to the best of our knowledge are selectivities considerably larger than those obtained with synthetic membranes reported previously but in line with selectivities dealt with in solvent-extraction separations.

In Table 2 we see uranium fluxes and distribution coefficients through a series of similar extracting solvents, namely, the series of mixed butyl-cresyl phosphates. This series was investigated in order to check the possibility of increasing the lifetime of the membrane by improving the plasticizer properties of the extractant. It is interesting how permeability changes with the nature of the solvent and, primarily, of course, with the magnitude of K. We see that

TABLE 1. SELECTIVITY OF URANIUM WITH RESPECT TO OTHER NITRATES[a]

S^{I}	S^{II}	$J_u^o/C_u^I \times 10^6$, cm/sec	$J_m^o/C_m^I \times 10^6$, cm/sec	β	K_u
0.02M $UO_2(NO_3)_2$ 3.0M HNO_3	H_2O	80	32	2.8	60
0.02M $UO_2(NO_3)_2$ 3.0M $NaNO_3$	H_2O	150	2.5×10^{-3}	60×10^3	105
0.02M $UO_2(NO_3)_2$ 1.0M $Al(NO_3)_3$	H_2O	135	$>5 \times 10^{-3}$	$>27 \times 10^3$	110
0.02M $UO_2(NO_3)_2$ 1.0M $Fe(NO_3)_3$	H_2O	215	0.3	650	135

(a) Quoted from Reference (7).

Note: m in $J_m^o/C_m^I \times 10^6$ stands for metal cation of nitrate added to S^I.

TABLE 2. DIFFUSION COEFFICIENTS, D, BASED ON DISTRIBUTION COEFFICIENTS, K_u, BETWEEN AQUEOUS PHASE AND UNDILUTED PLASTICIZER(a)

Sample	J_u^o/c_u^I x 10^6	K_u 3M HNO_3	D
Triethylhexylphosphate	140	40	1.4×10^{-8}
Tributylphosphate	150	45	1.7×10^{-8}
Dibutylcresylphosphate	74	10	3.0×10^{-8}
Butyldicresylphosphate	40	0.6	2.6×10^{-7}
Tricresylphosphate	0	0	0

(a) Quoted from Reference (13).

the permeability changes from tributyl phosphate to dicresyl butyl phosphate by roughly a factor of 4, while K changes considerably more. Consequently, diffusion coefficients computed from these values vary more than expected. The data presented here show a rather interesting phenomenon, namely, the fact that the permeability of a membrane can be markedly changed by varying the functional groups bound to the membrane matrix. An even more drastic example of this phenomenon offers a comparison of a membrane of TBP with one of tributyl phosphine oxide where a rather small change in the nature of the solvent renders the membrane practically impermeable to uranium. It may be speculated that this is a possible way through which the permeability through biological membranes can be regulated quickly and drastically.

Looking again at Equation (2), we see that equilibrium is reached, i.e., permeant flow vanishes, when $K^I \cdot c^I = K^{II} \cdot c^{II}$. Provided that $K^I >> K^{II}$, large concentration phenomena are obtained based on the possibility of transporting the permeable species via a steep-gradient concentration uphill. This can be seen in Table 3, which shows the equilibrium distribution of uranium between water and sodium nitrate. The fact that the concentration ratio between compartments II and I can assume a value of 80 means that, by choosing a proper volume ratio, a remarkable concentration process in compartment II may be observed. Thus equilibration of the complex between the two different solutions facing the membrane is independent of transport parameters, attaining the same maximal concentration effect as would be reached by solvent extraction of the permeant from compartment I to compartment II.

The uneven distribution of uranyl nitrate between two phases separated by a solubility membrane resembles Donnan distributions as commonly demonstrated with dialysis and ion-exchange membranes. In both cases, the activity of the permeable species is determined by the activity of a component that is restricted to one phase. In analogy to the Donnan equilibrium, the equilibrium condition can be obtained as the distribution where the activity of the permeable uranyl nitrate complex is equal in both compartments:

$$a_u^I \left(a_{NO_3}^I\right)^2 = a_u^{II} \left(a_{NO_3}^{II}\right)^2 \qquad (3a)$$

TABLE 3. EQUILIBRIUM DISTRIBUTION OF URANIUM BETWEEN WATER AND SODIUM NITRATE[a]

	c_{NaNO_3}		c_u	K_u	c_u^{II}/c_u^{I}		K_u^{I}/K_u^{II}	
(1)	S^I	2.0M	$8.7\cdot10^{-3}$M	64	22.8	0.8	23.7	1.3
	S^{II}	--	0.198M	2.7				
(2)	S^I	3.0M	$2.4\cdot10^{-3}$M	224	84	3	83	5
	S^{II}	--	0.202M	2.7				

(a) Quoted from Reference (7).

Activities of the permeable uranium complex in this equation can be replaced by the stoichiometric concentrations of uranium and its activity coefficients f_u^I, f_u^{II}, $f_{HNO_3}{}^I$, and $f_{HNO_3}{}^{II}$. The coefficients f_u^I and f_u^{II} also contain the complex formation constant of the uranium complex:

$$\frac{c_u^I}{c_u^{II}} = \frac{f_u^{II}}{f_u^I} \cdot \frac{\left(f^{II}NO_3\right)^2}{\left(f^{I}NO_3\right)^2} \cdot \frac{\left(c^{II}NO_3\right)^2}{\left(c^{I}NO_3\right)^2} = F \cdot \frac{\left(c^{II}NO_3\right)^2}{\left(c^{I}NO_3\right)^2} \qquad (3b)$$

Unequality of either c_u^{II}/c_u^{I} or K_u^{II}/K_u^{I} with $(c^INO_3)^2/(c^{II}NO_3)^2$ are due to deviations of F from unity.

As the purpose of this lecture is to stress phenomena of general interest encountered specifically in solvent membranes, the permeability of which is based on metal complexation, we shall mention two rather unexpected phenomena. The first deals with the kinetics of distribution of a permeable species through a complexing membrane; the second is encountered with composite membranes consisting of a thin, nonporous solvent membrane reinforced in series by a porous support.

With respect to the first phenomenon, the previously described concentration effect of the permeant in S^{II}, i.e., at the side of the membrane with a lower distribution coefficient, can only be achieved if the metal ligand that causes K to be large in S^I is completely impermeable. In any other case, e.g., if the nitrate in S^I is present in the form of nitric acid, which has a lower but appreciable permeability, uranium accumulates in S^{II} as long as little nitric acid has permeated and K^{II} is small. With increasing concentration of nitric acid in S^{II}, K^{II} increases and causes back diffusion of uranium. This overshooting phenomenon is shown in Table 4, where the concentration in S^{II} is shown as a function of time. We see that, under conditions in which the metal ligand permeates, uranium concentration in S^{II} overshoots; that is, after an initial accumulation decreases and approaches an even distribution between S^I and S^{II}. We have mentioned the phenomenon for two reasons: first, it may be of general interest; and second, to

TABLE 4. DISTRIBUTION OF A GIVEN AMOUNT OF URANIUM NITRATE BETWEEN TWO COMPARTMENTS, S^I, CONTAINING NITRIC ACID, AND S^{II}, INITIALLY CONTAINING PURE WATER[(a)]

Initial concentration of nitric acid in S^I – 4.7 mole.

Time, hr	S_u^{II}	S_u^I	S_u^I/S_u^{II}	$S^{II}HNO_3$	S^IHNO_3	S_H^{II}/S_H^I
4	2.8×10^{-3}m	4.2×10^{-3}m	0.66	0.4M	4.3M	0.09
10	4.9×10^{-3}m	0.9×10^{-3}m	6.1	0.9M	3.8M	0.24
20	3.6×10^{-3}m	2.0×10^{-3}m	1.8	1.6M	3.0M	0.53
120	3.0×10^{-3}m	2.6×10^{-3}m	1.1	2.1M	2.1M	0.86
240	2.8×10^{-3}m	2.8×10^{-3}m	1.0	2.3M	2.3M	1.0

(a) Quoted from Reference (13).

show that the use of permeable ligand in S^I creates difficulties which for practical applications of solvent membranes for metal complexes have to be overcome.

A second phenomenon worth mentioning is encountered with composite membranes. The consideration outlined so far referred to a membrane constructed of a single element. Since reasonable rates of transport can be achieved only by thin and therefore mechanically weak membranes, it is often necessary to reinforce the active film by an inert, porous support. For example, a high-wet-strength paper was used for this purpose.

In order to analyze the influence of the support on the performance of the composite membrane, we first consider, for simplification, the flow of uranyl nitrate in the presence of a nonpermeating nitrate species such as sodium nitrate.

The flow of uranyl ions through the paper, $J_{u,p}$, is given by Fick's law[(7)]:

$$J_{u,p} = P_p\left(c_u' - c_u^{II}\right) \tag{4}$$

where c' is the concentration of uranyl nitrate at the surface of the paper adjacent to the coating and

$$P_p = \frac{D_u^o \cdot a_p}{d_p} ,$$

with D_u^o being the diffusion constant of uranyl nitrate in water (assumed as approximately constant), a_p the available area per cm^2, and d_p the thickness of the paper. Flow through the coating, $J_{u,c}$, is given by

$$J_{u,c} = \frac{D_{u,c}}{d_c}\left(K^I_u c^I_u - K'_u c'_u\right) = P^I_c\left(c^I_u - K'_u/K^I_u \cdot c'_u\right) \tag{5}$$

where $D_{u,c}$ denotes the diffusion coefficient in the coating, d_c the thickness of the coating, K'_u the distribution coefficient of uranyl nitrate between the coating and the aqueous phase at the surface of the paper adjacent to the coating, and P^I_c is defined as

$$P^I_c = \frac{D_u{}^o \cdot K^I_u}{d_c}$$

From Equations (4) and (5) we obtain the concentration at the interface in the stationary state (with the uranyl nitrate-containing solution facing the **paper** side of the membrane):

$$c'_u = \frac{c^I_u + c^{II}_u \cdot P_p/P^I_c}{P_p/P^I_c + K'_u/K^I_u} \quad . \tag{6a}$$

Equation (6a) shows clearly that, during permeation into a large, stirred compartment, where $c^{II}_u \approx 0$, c'_u may be larger than c^I_u, provided that

$$P_p/P^I_c + K'_u/K^I_u < 1 \quad .$$

This is the case in our supported membrane, where P_p is much smaller than P^I_c (Table 5) and K'_u is smaller than K^I_u in the absence of excess nitrate ion at the coating/paper interface. Whenever $c'_u > c^I_u$, flow of uranyl nitrate through the coated paper will be larger than flow through the paper alone, i.e., **the presence of the coating is expected to enhance permeation through the paper**. The concentration profiles through paper alone and through a composite membrane are shown schematically in Figure 5.

In Table 5 are listed uranium fluxes through paper and coating as well as through coated paper. It is seen that the permeability through paper is increased five-fold by the presence of the coating, confirming the conclusions from Eq. (6a). The acceleration of uranium transport through the paper layer indicates that the concentration gradient of uranium across the paper layer is larger in a composite membrane than in paper alone:

$$\left(c'_u - c^{II}_u\right) > \left(c^I_u - c^{II}_u\right) \quad .$$

The accumulation of uranium in the unstirred region just adjoining the coating is caused by K'_u being smaller than K^I_u.

According to Eq. (6a), for **equal** concentrations of nitric acid at both sides of the membrane (i.e., $K'_u \approx K^I_u$), c'_u must be smaller than c^I_u, since P_p is much larger than P^I_c.

TABLE 5. INFLUENCE OF PAPER SUPPORT[a]

Membrane	S^I	S^{II}	Uranium Permeability Coefficient x 10^6, cm/sec
Coating[b]	0.02M $UO_2(NO_3)_2$ 3M HNO_3	H_2O	$P_c = 240$
Paper	0.02M $UO_2(NO_3)_2$ 3M HNO_3	H_2O	$P_p = 20.9$
Coated paper, coating exposed to S^{II}	0.02M $UO_2(NO_3)_2$ 3M HNO_3	3M UNO_3	$J_u^o/c_u^I = 15.7$
Coating exposed to S^I	0.02M $UO_2(NO_3)_2$ 3M HNO_3	3M HNO_3	$J_u^o/c_u^I = 110$
Coating exposed to S^I	H_2O	0.02M $UO_2(NO_3)_2$ 3M HNO_3	$J_u^o/c_u^{II} = 31.4$

(a) Quoted from Reference (7).
(b) Extrapolated to weight of coating in coated paper.

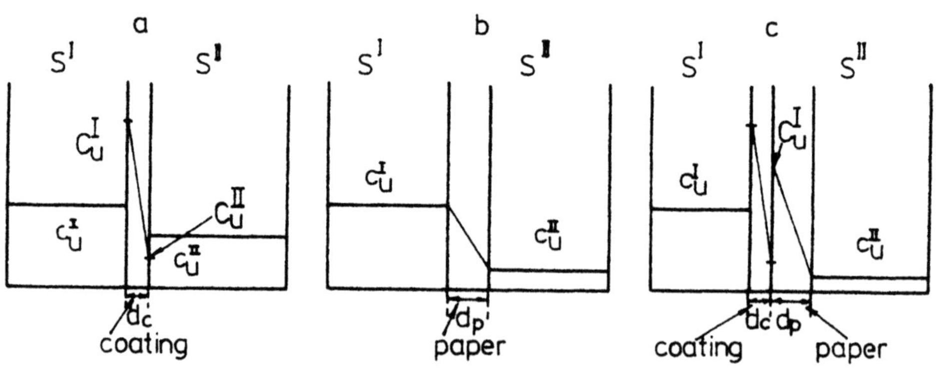

Figure 5. Concentration profiles through differing media.(*)

a. Unsupported membrane (coating).
b. Paper support.
c. Composite membrane of coated paper.
(*) Quoted from Reference (7).

Permeation through the composite membrane thus should be less than that through paper alone. This was confirmed by the data from Table 5 (see center section of data).

To summarize this phenomenon, *the combination of a solvent membrane with a porous membrane can, under favorable conditions, result in a composite system in which the permeation resistance of the paper is actually decreased by the presence of the coating. The coating enhances the diffusion through the paper.*

It should be noticed, however, that a membrane process has inherent disadvantages compared with a solvent-extraction process. In solvent extraction a multistage process usually is required, multiplying single selectivities manyfold. Such a possibility is not feasible in the case of a membrane process, where permeability rates are so slow as not to allow multiple transfer.

In order to be practical, therefore, a membrane process has to be based on an appreciably higher selectivity than encountered in solvent extraction. Recent experiments show (in fact, promise) that solvent-type membranes that originally were prepared to reach the selectivity encountered in solvent extraction undoubtedly can exceed those selectivities.

Expectations that membrane permeation through complexing membranes and solvent extraction would be processes of comparable selectivity are based on the assumption that the magnitude of the permeability coefficient, defined previously as

$$P = \frac{K \cdot D}{d} \quad , \tag{1}$$

is essentially determined by K, which may vary over many orders of magnitude while the membrane-diffusion coefficient, D, does not vary within too wide a range. This assumption would seem justified if the possible range of variations of diffusion coefficients in liquids is taken into account. Such an analogy, however, may not be correct. Both earlier and relatively new experimental evidence obtained by the various research teams in Israel active in this field (as well as theoretical considerations by Eigen) seem to indicate that membrane permeabilities may vary considerably more than the respective distribution coefficients.

Eigen[10], within the framework of his studies on complexation kinetics and, in particular, its implication on biological phenomena, has made two statements of interest in the present context:

(1) Complexation rate constants vary considerably more than complex stability constants

(2) As a consequence, membranes (the diffusion of which is based on a complexation process) may show surprisingly large differences in their permeabilities of chemically rather similar species.

Eigen, who was primarily concerned in giving a working hypothesis to explain permeability differences through liquid membranes, estimated – admittedly, by making some rather daring assumptions – the following differences in dissociation rates of the alkali complexes of uranyl diacetic acid:

Li^+	$5 \cdot 10^3$ sec^{-1}
Na^+	10^6 sec^{-1}
K^+	$2 \cdot 10^7$ sec^{-1}

He points out that these dissociation rates, translated into membrane permeabilities, yield extraordinarily high selectivities. Lacking techniques to incorporate uranyl diacetic acid into a membrane matrix, Eigen did not try to demonstrate these differences in complexation rates experimentally as differences of permeabilities. He speculated, however, that the kinetics of carrier transport through biological membranes may be determined by this effect.

In order to evaluate the increased selectivity of such membranes, we have to estimate the expected effect of the differences in complexation rates on the selectivity of solvent-type membranes. If we assume Eyring's diffusion model(11) to be valid in a solvent-type membrane, then the influence of the complexation rate constant may be approximated. Eyring presented a simple relationship between the diffusion coefficient, D, and the formation or dissociation rate of the associate of the diffusing species with the activated sites on its diffusion path,

$$D = \frac{\lambda^2}{2\tau} \tag{7}$$

where λ denotes the distance between two reactive sites and τ is the relaxation time or the reciprocal reaction constant determining formation or dissociation of the associate. The application of this relation to membrane permeation is based on the model that transport of a permeating species is based on the jumping rate from one reactive site of the membrane matrix to the next.

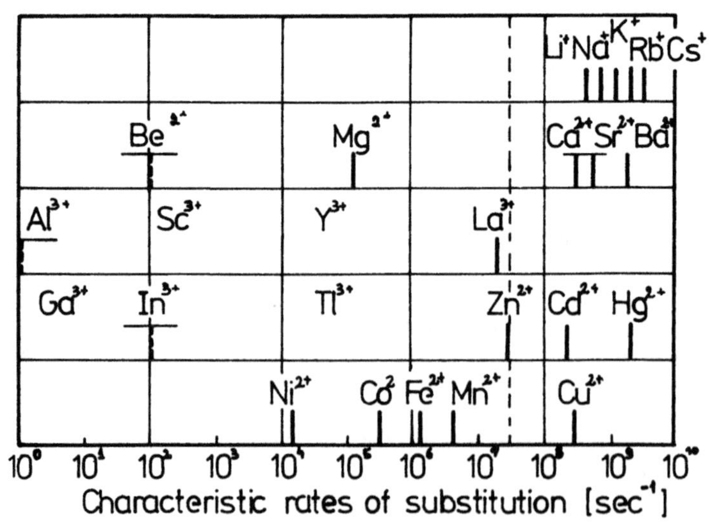

Figure 6. Characteristic rate constants for H_2O substitution in the inner coordination sphere of metal ions.(7)

It is known that chemical reaction rates vary over a wide range – according to Eigen(12), from 10^2 to 10^{-14} sec^{-1} (Figure 6). Few reliable data are known; a well examined reaction is the substitution of water from the hydration shell of different metal ions, where reaction rates vary over 10 orders of magnitude. In a general reaction, the rate constants of two complexes may differ considerably more than their equilibrium constants. It is therefore to be anticipated that a membrane process based on differences in complexation rates will be very much more selective than a solvent-extraction process based on differences of the respective equilibrium constants.

Thus, to tailor make membranes of extraordinary selectivity between two metal ions A and B, a chelating agent should be chosen which is known to show a large difference in chelation rate between these ions. Unfortunately, complexation kinetics is a rather virgin field and no suitable data have been found to construct membranes the selectivity of which could be predicted on the basis of complexation rates.

The experimental evidence to support the model to obtain highly selective membranes based on difference in complexation rates is still scant.

Table 6 shows permeabilities, distribution coefficients, and diffusion coefficients of a number of plasticizers which are selective extractants of the uranyl ion UO_2^{++}. It is readily observed that, while the distribution coefficient does not vary over too wide a range, the permeability or diffusion coefficient changes over at least 4 orders of magnitude.

Lacking any other explanation, we attribute this phenomenon to varying rates of complexation. It is readily seen that the plasticizers containing high distribution coefficients with low permeabilities are polyfunctional plasticizers containing the P=0 group at least twice per

TABLE 6. PERMEATION RATES, DISTRIBUTION COEFFICIENTS, AND COMPARATIVE PERMEABILITIES FOR DIFFERENT PLASTICIZERS(a)

Sample	$J_u^O/c_u^I \cdot 10^6$	K_u 3M HNO_3	D
Tributylphosphate (TBP)	150	9.2	1.3×10^{-8}
Triethylhexyl phosphate (TEHP)	140	6.0	2.3×10^{-8}
Tricyclohexyl phosphate (TCHP)	0.5	10.5	3.8×10^{-11}
Tributyl phosphine oxide (TBPO)	$0.05 \cdot 10^{-2}$	28	1.4×10^{-12}
o-Tris(dibutyl-phosphonopropyl) phosphate	0.15	10.0	4×10^{-11}
Bis(3-dibutyl phosphonopropyl) adipate	$.05 \cdot 10^{-2}$	17.4	2.3×10^{-12}
Polyvinyl dibutyl phosphonate (PVDBP)	$.07 \cdot 10^{-2}$	7.7	7.3×10^{-12}
Tetrabutyl ethylene diphosphonate (TBEDP)	1.6	27.6	4.6×10^{-11}
Dibutylcresyl phosphate (DBCP)	74	0.6	1.0×10^{-7}
Butyl dicresyl phosphate (BDCP)	39	0.03	1.0×10^{-6}
Tricresyl phosphate (TCP)	0	0	0

(a) Quoted from Reference (13).

molecule. This leads one to assume that the resulting uranyl complex is a chelate, a type of complex known to show considerably lower reaction rates than the ion-association complexes obtained with monofunctional phosphate plasticizers.

While the experiments summarized in Table 6 show different permeabilities of the same ion through different but similar membranes, we can also show permeabilities of different ions through the same membrane. To demonstrate this phenomenon with as large as possible an effect, a membrane was prepared containing an immobilized chelating agent known to complex slowly, namely, thenoyl trifluoroacetone. Permeabilities and distribution coefficients of this membrane for ferric, cupric and zinc ions were determined, and, from these data, the diffusion coefficients were calculated. The diffusion coefficients change as shown in Table 7.

The great differences between the permeabilities of ferric and cupric ions promise easy separation of these ions by selective dialysis. A separation experiment between ferric ions and cupric ions was carried out at a pH giving equal distribution coefficients. It resulted in a clear-cut separation of the two ions. For reasons not yet understood, the difference in diffusion coefficients between the two ions was even higher in the case where both ions diffused simultaneously through the membranes. The 200-times-larger diffusion coefficient of cupric ion as compared with ferric ion is a promising demonstration of the potential of membranes with permeability based on reaction rates for separations.

Thus, in two independent sets of experiments – Eigen's computed diffusion times of alkali ions through the cell wall and our permeabilities of d-elements through chelating membranes – seem to demonstrate the existence of a selective principle to which hitherto not enough attention has been paid. It may well be that significant effects could be observed in the case of very similar transition metals, and we are, in fact, about to check whether such effects can be observed in the case of metals which are very difficult to separate by conventional chemical means, such as rare earth metals or other f-elements.

The concept that reaction rates may be decisive in determining permeability rates is not restricted to complexing membranes only. In the context of our work on reverse-osmosis membranes, we have observed separately that water content and the rate of water transport may be completely unrelated. This phenomenon is sometimes accounted for by assuming the existence of bound and free water or, alternatively, by the concept of water clusters. It may be, however, that the rate-determining step is the rate of hydrate formation or dissociation. Similar

TABLE 7. PERMEABILITY, DISTRIBUTION AND DIFFUSION COEFFICIENTS OF FERRIC, CUPRIC, AND ZINC IONS WITH A MEMBRANE CONTAINING IMMOBILIZED THENOYLTRIFLUOROACETONE

	P (cm/sec)	K	D_{abs}	D_{rel}
Fe^{3+}	$1.7 \cdot 10^{-6}$	4.4	$5 \cdot 10^{-9}$	1
Zn^{2+}	$7.6 \cdot 10^{-6}$	5.3	$26 \cdot 10^{-9}$	5
Cu^{2+}	$23.2 \cdot 10^{-6}$	4.2	$70.4 \cdot 10^{-9}$	15
Cu^{2+} } simultaneously	--	--	--	1
Fe^{3+} } simultaneously	--	--	--	200

phenomena are observed in liquid permeations or separations of gas and could conceivably be explained by the same concept. Current efforts to measure such rate processes directly by various spectroscopic methods may substantiate or repudiate this theory.

To conclude, membrane processes, as is clearly evident from the program of this meeting, are being followed with ever growing interest for their practical applications. This development is paralleled by a trend to apply solvent extraction as a hydrometallurgical process to ever cheaper materials. We tried inter alia to show here that the two processes may very well merge. We are, in fact, fairly hopeful that the wealth of technical innovations developed mainly for the process of reverse-osmosis desalination may soon shift solvent-membrane dialysis as a hydrometallurgical unit operation into the range of economic feasibility.

REFERENCES

(1) E. Overton, *Pfluegers Arch. ges. Physiol.*, **92**, 115 (1902).

(2) B. F. Warner, *Solvent Extraction Chemistry,* North-Holland Publishing Co., p 635 (1965).

(3) J. H. Schulman, and H. L. Rosano, in *Retardation of Evaporation by Monolayers,* by V. La Mer, Academic Press, p 97 (1962).

(4) K. Sollner, and G. M. Shean, *J. Am. Chem. Soc.*, **86**, 1901 (1964).

(5) A. Ilani, *Biophysical Journal,* **8**, 556 (1969).

(6) R. Bloch, O. Kedem, and D. Vofsi, *Solvent Extraction Chemistry,* North-Holland Pub. Co., p 605 (1965).

(7) R. Bloch, A. Finkelstein, O. Kedem, and D. Vofsi, I. and E.C. *Process Design and Development,* **6**, 231 (1967).

(8) Bankman and Z. Specker, *Anal. Chem.*, **162**, 18 (1958).

(9) H.A.C. McKay, *Chem. and Ind.*, **54**, 1545 (1954).

(10) M. Eigen, *Pure Appl. Chem.*, **6**, 105 (1963).

(11) H. Eyring, *J. Chem. Phys.*, **4**, 283 (1936).

(12) M. Eigen, and R. D. Wilkins, *Mechanism of Inorg. Reactions,* Am. Chem. Soc. Monography, Washington, D. C. (1965).

(13) R. Bloch, A. Finkelstein, S. Marian, D. Vofsi, and O. Kedem, Plastics Research Laboratory Report No. 65-1.

LOW-PRESSURE ULTRAFILTRATION SYSTEMS FOR WASTEWATER CONTAMINANT REMOVAL

Liquid Membrane Studies

W. Leigh Short, Rolf T. Skrinde, and Donald G. Newton, Jr.*

INTRODUCTION

Low-pressure ultrafiltration is still in the preliminary stages of development as a unit process in municipal waste and wastewater treatment. Recently, the membrane process of reverse osmosis, developed initially for use in desalination, has attracted attention as a promising method for treatment of wastewaters. Data on the separation or removal of pure organic solutes is needed to evaluate the potential application of ultrafiltration systems in wastewater treatment. The organic solutes in wastewater streams cannot, in general, be economically recovered by chemical means. Ultrafiltration has the potential of making possible the economic treatment of at least some industrial wastewaters. A pertinent example is the pulp and paper industry, in particular, research efforts now being conducted by the Pulp Manufacturers Research League.(1,2,3) Studies carried out on reverse-osmosis treatment of secondary-sewage plant effluent at Pomona, California(4), have demonstrated the feasibility of reverse osmosis for treatment of complex wastewaters containing relatively high concentrations of organic pollutants.

This paper presents data obtained at the University of Massachusetts on removal of organic material from water by low-pressure ultrafiltration in the presence of a surface active agent. The organic material used was raffinose, and the surface active agents used are listed in Table 1. Additional data taken in our laboratories on the use of low-pressure ultrafiltration to separate organic material from water is presented elsewhere.(5)

The use of a surface active agent to increase the salt rejection of a cellulose acetate membrane apparently was encountered first by Martin.(6) The addition of a small amount of

*Respectively: Associate Professor of Chemical Engineering, Professor of Civil Engineering, and Graduate Student studying for the M.S. Degree in Environmental Engineering at the University of Massachusetts. Dr. Skrinde became Chairman of the Department of Civil Engineering at the University of Iowa in September, 1969.

TABLE 1. SURFACTANTS USED IN LIQUID MEMBRANE STUDIES

Trade Name	HLB[a] ±1	Chemical Name
Tween - 20	16.7	Polyoxyethylene (20) sorbitan monolaurate
Myrj - 52$	16.9	Polyoxyethylene (20) stearate
Brij - 58	15.7	Polyoxyethylene (20) cety ether
Igepac Co - 850	--	Nonylphenoxypoly (ethyleneoxy) ethanol N=20
Polyox WSR - 301	--	Polyoxyethylene polymer

(a) HLB = Hydrophile/lipophile balance.

polyvinyl methyl ether (PVM) to the saline feed solution during ultrafiltration resulted in a reduction in the transport of solute across the membrane while at the same time causing a slight decrease in the product-water permeation rate.

Michels, et al.[7], investigated the use of PVM with leaky desalination membranes. They reported that PVM treatment significantly reduced the salt flux at any pressure without significantly altering the product-water flux. They also reported that PVM treatment reduced the pressure dependency of the salt flux. They hypothesized that the PVM restricted or blocked the flow of saline solution through "micropores" in the polymer matrix that might exist in the membrane but in no way interfered with diffusive transport of ions through the rest of the membrane; i.e., they assumed that salt flux occurs by ion diffusion in the membrane matrix and not by diffusion through water-filled pores.

Kesting[8] more recently has studied effects of surface active agents on salt rejection by cellulose acetate membranes. He postulated that a layer of surfactant forms at the saline solution/cellulose acetate membrane interface. At low concentrations this layer is incomplete, but it finally covers the membrane surface when the critical micelle concentration (CMC) is reached. At the CMC, all surfaces are covered by a very thin layer of surfactant. A significant increase in surfactant concentration beyond the CMC would result in the surfactant aggregating into micelles in the bulk solution.[9]

Kesting found that the major reduction in solute transfer occurred before the CMC was reached in the saline solution. At the CMC, the surface layer was assumed to be fully developed, and he attributed any changes in salt flux above the CMC to compaction of the surfactant layer as more molecules tried to become oriented at the surface. Kesting also assumed that the surfactant layer was permselective; i.e., it acted to impede transport in the same manner as a solid membrane.

Kesting reported considerable variation among the surface active agents. He concluded that those surfactants which possessed a high hydrophile-lipophile balance (HLB) provided the greatest salt-rejection efficiency while at the same time presenting the least resistance to pure-water transport.

The purpose of this study was to evaluate the effect of surfactants during low-pressure ultrafiltration of dilute raffinose solutions. The surfactants selected are listed in Table 1; all have high HLB ratios.

EXPERIMENTAL APPARATUS

The low-pressure ultrafiltration apparatus is shown in Figure 1. The system is designed to operate in the 20 to 90-psi range at temperatures between 20 and 70 C. Three Dorr Oliver-type low-pressure ultrafiltration cells (0.05 of a square foot of membrane surface, each) are operated in parallel using feed from a single reservoir. The permeate is recycled so that the reservoir concentration is held constant (except, of course, for minute changes resulting from the 1-ml samples taken periodically; the reservoir capacity is 10.2 liters).

The surfactants were purchased from commercial suppliers and used to prepare 10 percent (by weight) solutions. The surfactants were heated to facilitate dispersal and the 10 wt % solutions were allowed to stand for 24 hours before use.

Abcor HFA-100 membranes were used in the Dorr Oliver cells. The pure-water permeation rates for this membrane, which are shown in Figure 2, were obtained using the low-pressure ultrafiltration apparatus.

ANALYTICAL PROCEDURES

The raffinose concentrations were determined using the colorimetric method described by Dubois.[10] The accuracy of the method is reported to be ±2 percent absolute. We conducted tests to show that the surfactants used did not interfere with the test procedure.

The absorbance of the sugar solutions was measured using a Bausch and Lomb Spectrometric 20 Colorimeter set at 490 mμ.

The surfactant concentrations were measured using a Perkin-Elmer 202 Ultraviolet-Visible Spectrophotometer set at 223 mμ and a slit width of 600.

RESULTS AND CONCLUSIONS

Table 2 tabulates the raffinose rejections (0.5 wt % raffinose) obtained at 40, 60, and 80-psi operating pressures and at surfactant active concentrations of 0 to 2500 mg/l. The temperature of operation was 25 C. As the table shows, the raffinose rejection is essentially independent of surfactant concentration and is, in fact, nearly constant at 80-85 percent. This agrees well with what would be expected using the HFA-100 membrane.[11] All of the surfactant concentrations are considerably above the CMC values reported for the surfactants used.[12,13] Therefore, a liquid membrane should have been fully developed at the membrane/sugar-solution interface. The surfactant, in addition to not altering the raffinose rejection, had no discernible effect on the permeate flux.

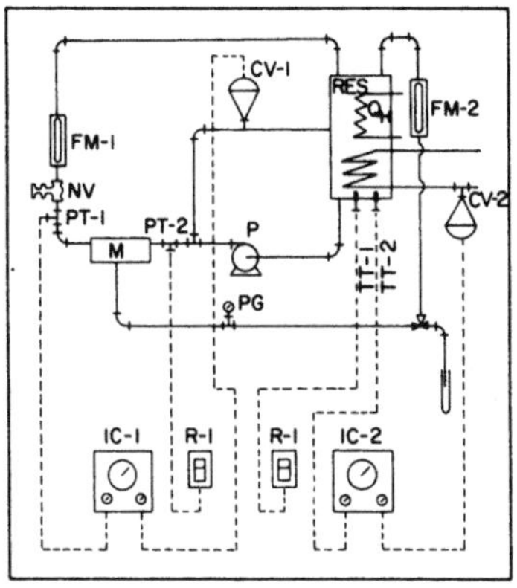

KEY FOR FIGURE 1

Symbol	Explanation	Symbol	Explanation
CV-1	Controller-operated pressure-control valve	P	Double-lobe rotary pump
CV-2	Controller-operated temperature-control valve	PG	Pressure gage to measure back pressure on membrane
FM-1	Rotameter indicating flow across membrane (feed rate)	PT-1	Pressure transducer for IC-1
FM-2	Rotameter indicating flow through membrane (permeation rate)	PT-2	Pressure transducer for R-1
IC-1	Indicating controller – outlet pressure for reverse-osmosis cell (M)	RES	Reservoir for solute-solvent system
IC-2	Indicating controller – system temperature	R-1	Inlet-pressure recorder for reverse-osmosis cell (M)
M	Dorr Oliver reverse-osmosis test cell (0.05 ft^2 approx. membrane area)	R-2	System-temperature recorder
NV	Needle valve to set feed rate across membrane surface	Q_H	Electric heating coil varied by means of a rheostat – to be used in conjunction with IC-2 for optimum controller action

Figure 1. Schematic of low-pressure ultrafiltration apparatus.

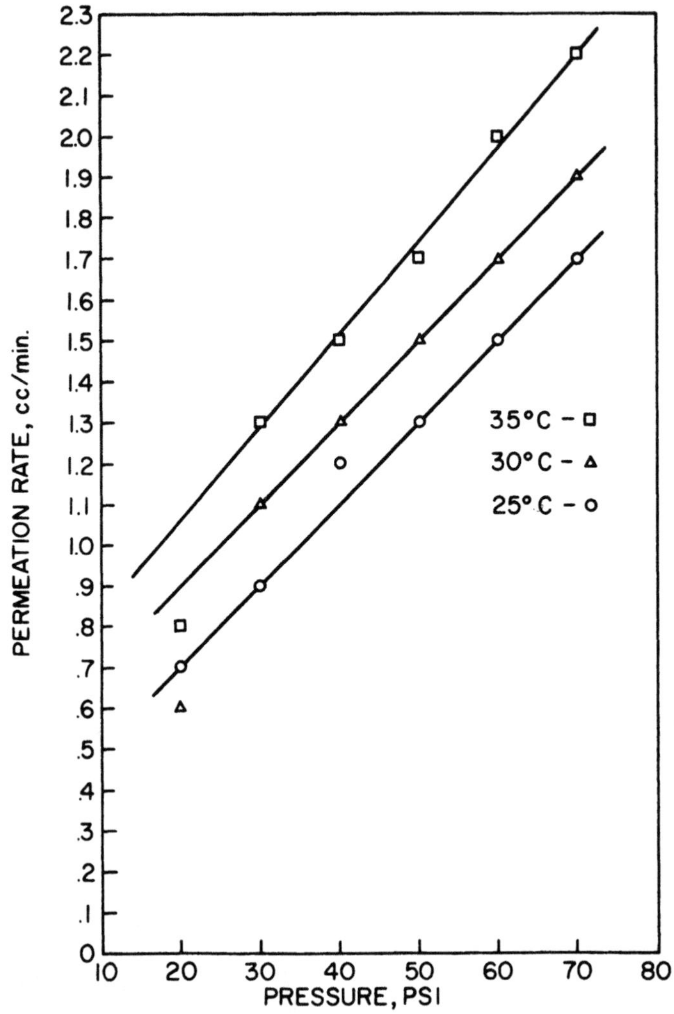

Figure 2. Effect of temperature on permeation rate, pure water, HFA 100-membrane.

Figure 3 shows the surfactant rejection efficiency at varying surfactant concentrations. The lower line, which shows rejections of approximately 40 percent, is for pure surfactant (no raffinose); the middle line shows the surfactant rejection that occurs at various surfactant concentrations when 0.05 wt % raffinose is present. The upper curve indicates that the rejection of the 0.05 wt % raffinose in the presence of various surfactant concentrations remains constant at about 84 percent.

The data shown in Figure 3 indicated that the surfactants used, which have a relatively high HLB ratio, did not significantly alter the raffinose rejection. This is an apparent contradiction to what would be expected from the work published by Kesting. For example, Kesting shows that the salt flux could be decreased by up to 50 percent while the water flux was decreased at the same time by approximately 50 percent.

Experimental studies have been reported on the effect of sucrose on spread monolayers at the air/water interface and on adsorbed filters.[14,15] With spread films of proteins and polyamine, the addition of sucrose to the aqueous substrate produced small changes in the surface pressure and the surface potential of the films at the interface. Marked effects were reported on the viscosity and elasticity which reflect changes in molecular bonding. No effect was reported on the surface properties of polyvinyl acetate and polyvinyl stearate. The authors concluded that sucrose exerts its effect by modifying the hydrogen bonding between keto-indo groups on neighboring chains.

TABLE 2. RESULTS IN PERCENT REJECTION (%R) OF LIQUID MEMBRANE STUDIES[(a)]

Membrane: ABCOR HFA-100
Feed Rate: 1 gpm

Concentration, mg/l	Pressure, Surface Active Agent					
	40 psi				60 psi	80 psi
	Tween-20	Myrj-52S	Igepac Co-850	Polyox	Brij-58	Brij-58
0	80	77	82	73	79	84
100	81			78	83	85
200				78	83	
300				79	83	
400				71	82	
500		79	81	73		
600				74		
700						
800						
900						
1000		81	83	78		
1100	79					
1200						
1300						
1400						
1500		82	83			
1600						
1700						
1800						
1900						
2000		81	83	79		
2100						
2200						
2300						
2400						
2500			84			

(a) Precision in %R = ±3.

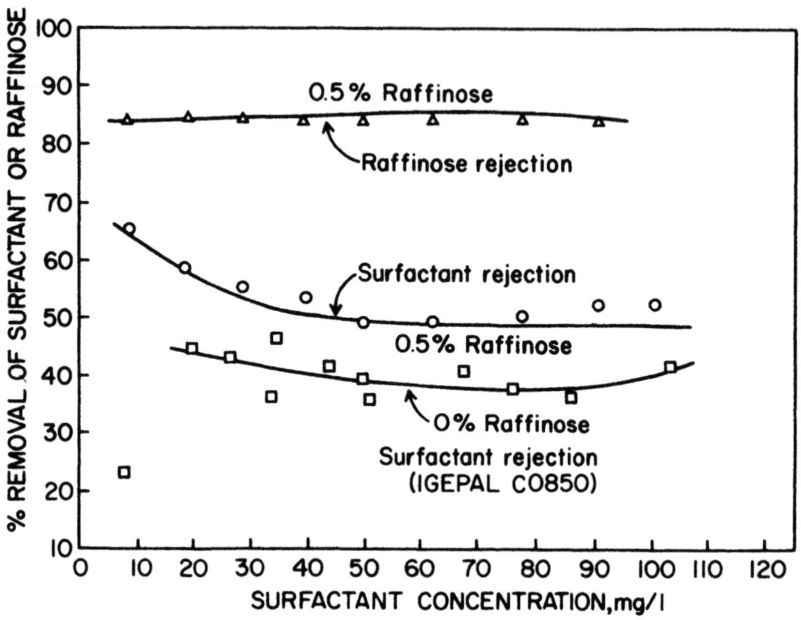

Figure 3. Surfactant rejection at various raffinose concentrations.

Taylor and Rawlinson(16) report that the hydrogen bonding of glucose and sucrose solutions is very strong, with sucrose having the larger number of hydrogen-bonding groups. We are now taking further experimental data, with sucrose, dextrose, and phenol as solutes, and using the same surfactants to form a liquid membrane. These data, which will be reported in the near future, should answer some of the questions implied in the above discussion.

ACKNOWLEDGMENT

One of the authors, Donald G. Newton, Jr., gratefully acknowledges the financial support of the Federal Water Pollution Control Administration through their grant Number 5T1-WP-77-04.

REFERENCES

(1) A.C.F. Ammerlan, and A. J. Wiley, "The Engineering Evaluation of Reverse Osmosis as a Method of Processing Spent Liquors of the Pulp and Paper Industry", paper presented at AIChE meeting, March, 1969 (New Orleans).

(2) A. J. Wiley, G. A. Dubey, J. H. Holderby, and A.C.F. Ammerlan, "Concentration of Dilute Pulping Wastes by Reverse Osmosis and Ultrafiltration", paper presented at WPCF meeting, October, 1969 (Dallas).

(3) A. J. Wiley, A.C.F. Ammerlan, and G. A. Dubey, *TAPPI*, **50**, 455-460 (1967).

(4) U. Merten, and D. T. Bray, "Reverse Osmosis for Water Reclamation", *Proceedings, Third International Conference on Water Pollution Research*, abstracted in *Jour. WPCF*, **38**, 397 (1966).

(5) R. T. Skrinde, W. L. Short, and D. J. Smola, "Reverse Osmosis Treatment of Wastewaters", to be published in *Journal of Applied Polymer Science*.

(6) F. Martin, reference number (1): private communication in Kesting (ref. 7).

(7) A. S. Michels, H. J. Bixler, and R. M. Hodges, Jr., "Kinetics of Water and Salt Transport in Cellulose Acetate Reverse Osmosis Desalination Membranes", *Jour. Colloid Science*, **20**, 1034-1056 (1965).

(8) R. E. Kesting, W. J. Subrasky, and J. D. Paton, "Liquid Membranes at the Cellulose Acetate Membrane/Saline Solution Interface in Reverse Osmosis", *Journal of Colloid and Interface Science*, **28** (1), 156-160 (1968).

(9) P. H. Elworthy, A. T. Florence, and C. B. MacFarlane, *Solubilization by Surface Active Agents*, Chapman Hall, Ltd., London (1968), Chapt. 1.

(10) M. Dubois, K. A. Gilles, J. K. Hamilton, P. A. Rebers, and F. Smith, "Colorimetric Method for Determination of Sugars and Related Substances", *Analytical Chem.*, **28** (3), 350-356 (1956).

(11) Private communication with ABCOR, Inc., October, 1969.

(12) J. J. Schick, S. M. Atlas, and F. R. Erich, "Micellar Structure of Non-Ionic Detergents", *JACS*, **66**, 1326-1333 (1962).

(13) P. Becher, "Micelle Formation in Aqueous and Non-Aqueous Solutions", *Non-Ionic Surfactants*, M. J. Schick, Ed., Marcel Debber, New York (1967), Chapt. 15.

(14) F. MacRitchie, and A. E. Alexander, "The Effect of Sucrose on Protein Films I. Spread Monolayers", *Jour. Colloid Science*, **16**, 15-61 (1961).

(15) F. MacRitchie, and A. E. Alexander, "The Effect of Sucrose on Protein Films II. Adsorbed Films", *Jour. Colloid Science*, **16**, 61-67 (1961).

(16) J. B. Taylor, and J. S. Rowlinson, "The Thermodynamic Properties of Aqueous Solutions of Glucose", *Trans. Faraday Soc.*, **51**, 1183-1192 (1955).

INDUSTRIAL WASTE TREATMENT OPPORTUNITIES FOR REVERSE OSMOSIS

J. G. Mahoney
Eastman Chemical Products, Inc.

M. E. Rowley
Polymer Technology Division

L. E. West
Photographic Technology Division

POSSIBLE INDUSTRIAL APPLICATIONS

Since the beginning of our industrial economy, an abundant supply of clean water has been a major factor in the choice of plant locations. In many instances in the past, industry has used water from our rivers and lakes, then returned it without regard to contaminants. Times are changing. Today's industrial leaders are more conscious of their responsibility to the public and are coming to regard adequate waste treatment as an integral part of the cost of doing business, and many are calling public attention to their efforts.

Still there are some companies, particularly those in fiercely competitive businesses, in which conventional waste treatment is regarded as a capital expense and operating cost that contributes nothing to the value of its products. Understandably, many have been reluctant to attempt to justify such expenditures to their stockholders and boards of directors without some overriding pressure from the outside.

Conventional primary and secondary waste treatment produce sludge which has to be disposed of, and an improved waste water that is generally discharged to a stream. Occasionally, the treated water might be satisfactory for reuse in some noncritical plant operation.

Much of what is called "industrial waste" is simply product, raw material, or potentially valuable by-product that process engineers have not been able to recover economically. It is for this reason that our publicity efforts for Eastman Membrane, describing a new chemical process for separating water from a dilute solution, have attracted the attention of process engineers in many industries. In the course of pursuing these inquiries, we have learned of a wide variety of applications where pollution abatement or waste treatment can also be a cost saving or even profit making operation when reverse osmosis or ultrafiltration is considered. You may be interested in hearing about a few of these inquiries. You also might be reminded of a similar situation in your own plant.

Processing wastes by reverse osmosis or ultrafiltration can bring about process economies in three ways:

(1) Recovery of the valuable materials dissolved in the waste stream

(2) Recovery of high quality water for reuse

(3) Concentration of pollutants for disposal.

Frequently, two or even all three of these benefits may apply in a particular situation. Of the following examples, the first two relate to a specific problem in a spacific plant. The potential cost savings that are mentioned came to light only because the companies faced a pollution problem. If savings such as these are possible, it might pay a lot of chemical companies to analyze minor waste streams and overflows even if there is no immediate need for improved pollution abatement practices.

Example 1. Recovery of Salts for Reuse

A chemical plant is producing in excess of designed capacity. The process effluent, containing sodium sulfate and sodium trithiocarbonate, flows to a holding tank, overflows at the rate of 2,000 to 3,000 gallons per day, and causes a pollution problem and a product loss of about $50.00 per day.

Although Eastman does not manufacture reverse-osmosis equipment, we could visualize a solution to the problem. The overflow could be fed to a reverse-osmosis unit using a flow rate and pressure to attain a 50-50 split of the overflow. The concentrate could be returned for reuse in the plant and the product water discharged to the sewer. Based on conversations we have had with equipment manufacturers, we think a total operating cost of $2.00 per 1000 gallons of product water is reasonable for a plant of this size. If the feasibility of this application is borne out in pilot studies, and the concentrate can be reused, then a $6.00 per day expenditure to solve a pollution problem will also result in a product recovery of $50.00 per day.

Example 2. Recovery and Reuse of Catalyst

A chemical plant recovers most of a cobalt catalyst used in its process stream. Despite its best efforts, there was still approximately 0.03 percent cobalt in the 10,000 gallons per day of waste water going to the "disposal ditch". Since the cobalt represents a serious pollution problem, the company is interested in a process to remove the cobalt from the effluent. Once again, reverse osmosis seems suited to the application. A 10,000-gallon-per-day plant, costing $20.00 per day to operate, will probably solve the pollution problem and permit the recovery of more than $60.00 per day in cobalt.

Example 3. Pulp-Mill Wastes

Processing of dilute pulp-mill wastes represents a potential application having industrywide possibilities. Most pulp mills already have adequate treatment methods for their cooking liquors, and many have processes to recover by-products from the solutions. Few, however, have been able to afford waste treatment facilities for the dilute waste streams. A pulp mill might discharge 1,000,000 gallons per day of solution having high biochemical oxygen demand (BOD), color, and odor, which do not meet evolving effluent standards. The Pulp Manufacturers Research League at Appleton, Wisconsin, under the direction of Mr. Averill Wiley, has been involved in an extensive program to demonstrate the feasibility of reverse osmosis to handle this problem. In several years of laboratory-scale testing, and now in a 100,000-gallon-per-day mobile demonstration unit, the League has demonstrated that dilute liquors containing 0.5 to 2 percent dissolved solids can be neutralized and then concentrated by reverse osmosis to the 8- to 12-percent range. The product water is practically color-free, odor-free, and nonfoaming. In a typical situation, well over 90 percent of the dissolved organic and inorganic materials present in the waste stream would be contained in the concentrate. BOD can be reduced as much as 95 percent.

Where are the offsetting economies in this pulp-mill application? First, for those mills that have a by-product recovery system from their cooking liquors, the reverse-osmosis concentrate can probably be added to the cooking liquor stream. Second, the purified water from the reverse-osmosis unit will in many cases be of better quality for pulp-mill purposes than the local surface or municipal supply and can be recycled to the most demanding section of the pulp mill. Third, for a mill that does not have a by-product recovery system, yet is still faced with a deadline for installing treatment facilities, reverse osmosis could reduce the volume of effluent from 1,000,000 gallons per day to 200,000 gallons or less. Thus, the capital cost for an evaporation and combustion system could be substantially reduced.

Example 4. Food Industry

The food industry would like to recover sugars, starches, and proteins dissolved in effluent streams. Many of these are being covered in other papers here by our friends from the U. S. Department of Agriculture, who have done most of the investigations.

Example 5. Removal of Dyes

Many dyestuffs are produced by "salting out" from an aqueous solution. Some dyestuff manufacturers have been discharging the spent solution directly to surface waters or municipal sewers, but the persistent color constitutes an aesthetic problem regardless of whether or not it constitutes a pollution problem. One manufacturer in the East was under pressure to remove the color from his effluent. By means of reverse osmosis, he should be able to achieve an 80- to 90-percent reduction in volume of colored solution. This should greatly reduce the capital cost of any subsequent evaporation or combustion system that he might need to install, and might also permit additional dye recovery from the more concentrated solution.

Example 6. Car-Wash Operations

It is sometimes desirable to install an automatic car wash at a location that does not have access to municipal sewers. The waste water may not be acceptable for discharge into a storm sewer or surface stream. Where this condition exists, the only answer is to remove the waste water by truck. Some car-wash installations now utilize demineralized water for a nonspotting final rinse. In the situation described here, reverse osmosis could reduce the cost of waste disposal by concentrating the waste by a factor of 5 or 6; and, at the same time, the purified water can be used for the final rinse, saving the cost of water that would otherwise have to be bought from the municipality and purified by ion exchange.

Up to this point, we have been reporting only on what other companies are considering. Now we will turn to some experimental work that has been done in the Eastman Kodak Company. Dr. Gene Rowley will describe these studies on the recovery and reuse of chemicals used in the processing of photographic film and paper. These initial studies were to evaluate the technical feasibility of recovery and reuse of chemicals in photographic systems being discharged from the processes. Economic evaluation will follow the most promising leads.

APPLICATIONS TO PHOTOGRAPHIC PROCESSING

Introduction

Photographic processing is under study as a possible application for reverse osmosis. There is large-scale photographic processing in all metropolitan areas and many suburban areas of the United States. The customers' film is spliced together and processed on a machine as a continuous strand. The film passes through several processing solutions that are replenished continuously. The processing solutions in the tanks overflow at the same rate as they are replenished. Each processing solution is generally followed by a tank of wash water. A simplified machine is diagrammed in Figure 1. Some of the processing solutions are carried over by the film into the succeeding wash water. The wash-water flow rate is generally from 4 to 20 times that of the total replenisher rates of the processing solutions in order to remove the unwanted processing chemicals from the film. Thus the wash waters are very dilute compared to the processing solutions.

It is from the dilute solutions that one might expect the most favorable results by using reverse osmosis. If the chemicals in the wash waters could be recovered and reused, this would (1) reduce the amount of processing chemicals that must be purchased, (2) produce product water that might be reused for washing, and (3) reduce the chemicals that are discharged to the sewer.

It should be emphasized that the experiments described below constitute a progress report on a research study of possible applications of reverse osmosis to photographic processing. None of these have yet been confirmed in a commercial trial.

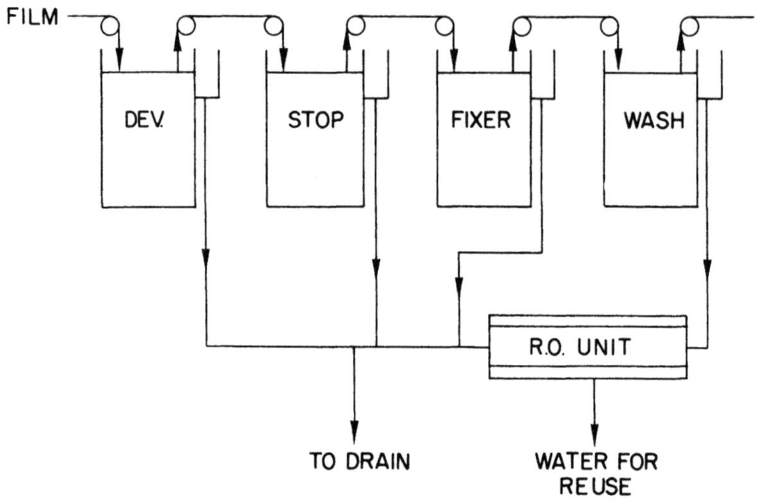

Figure 1. Continuous photographic processing machine with reverse-osmosis unit.

Procedure for Research Experiments

Laboratory Cell. The flow diagram for the laboratory test unit is shown in Figure 2. The cell accepts a membrane disc with an effective area of 2.07 in^2. The cellulose acetate membrane is backed with a filter paper followed by a sintered stainless-steel disc. The feed water at 75-85 F is delivered to the center of the membrane surface and forced radially to the circumference. The feed solution to be tested is pumped to the desired pressure and flow rate (maximum of 1,500 psi and 1,750 ml per minute). For more tests, 600 psi at 800-ml-per-minute flow was used.

The purpose of the unit was to study the separation of chemicals using various cellulose acetate membranes. Percentage water recovery was not of concern in the use of this test unit. The feed solution was held constant in all experiments. All data reported represent initial readings. No run exceeded 24 hours operating time.

Semiplant Equipment. Cellulose acetate membrane was coated on the inside of fiberglass tubes 8 feet x 0.5-inch ID.* They were connected in 10 or 20 modules of seven tubes each as shown in Figure 3. Each tube was connected in series within each module and the modules were connected in series with each other. Each module contained about 7 ft^2 of acetate membrane. The feed was pumped through the tubes, with the pressure of about 600 psi being controlled by a valve at the exit of the last tube. Shrouds surrounded each of the seven-tube

*These tubes were purchased with membrane material already coated on the inside by the vendor. Eastman Membrane was not used in these semiplant experiments but will be investigated in the future.

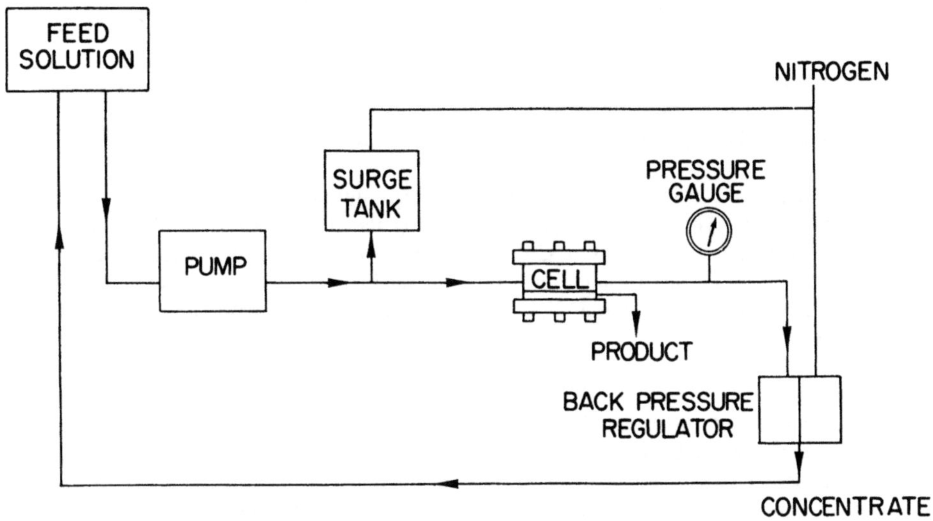

Figure 2. Laboratory test unit.

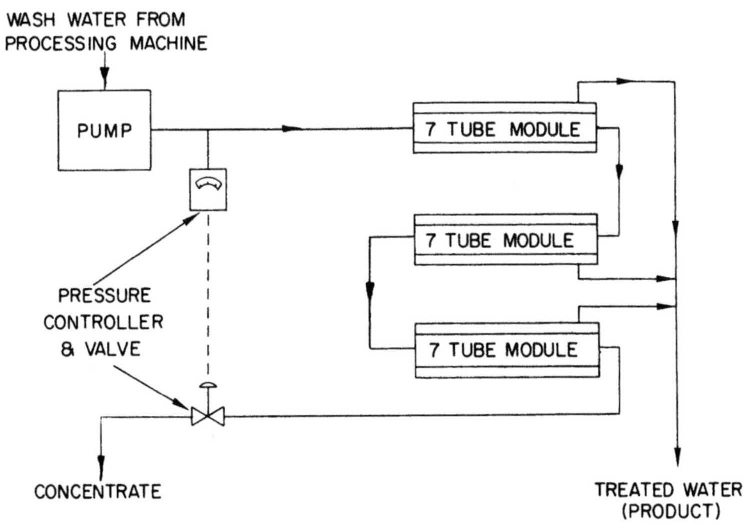

Figure 3. Reverse-osmosis unit with ten-module stack.

assemblies into which the product water passed. The product water collected in the shroud was at atmospheric pressure. The connections to a processing machine are shown in Figure 4.

Flux was dependent upon the particular ions present in the feed and on their concentrations. All modules in a given experiment yielded about the same flux and the same ion-rejection values. As would be expected, higher concentrations of feed caused lower flow rate of product.

A change was made in the equipment design during some runs to give a constant water-recovery rate. This was accomplished by fixing the concentrate flow rate and regulating the system pressure at the first module of the unit, overflowing any excess back into the solution holding tank prior to entry into the first module. Therefore the rate of the feed into the unit was not constant but was dictated by the fixed recovery rate and concentrate within the system.

Examination of Processing Chemicals. Some of the major chemicals used in processing have been examined individually to find the ones which could be separated most easily by reverse osmosis. The test results using the laboratory cell are shown in Table 1. The concentrations that were studied were those generally encountered in processing solutions or in wash waters. It is noted that the inorganic ions are rejected more completely than the organic molecules that were tested. Organic molecules are rejected in accordance with their size configuration and possibly by their relative absorption on the membrane[1].

Inorganic chemicals are used almost exclusively in photographic bleaches and fixing baths, thus those baths, or the washes that follow, are the most favorable for separation (to be concentrated) by reverse osmosis. Furthermore, it is especially desirable to recover ferro- and ferri-cyanides from bleach washes for reuse.

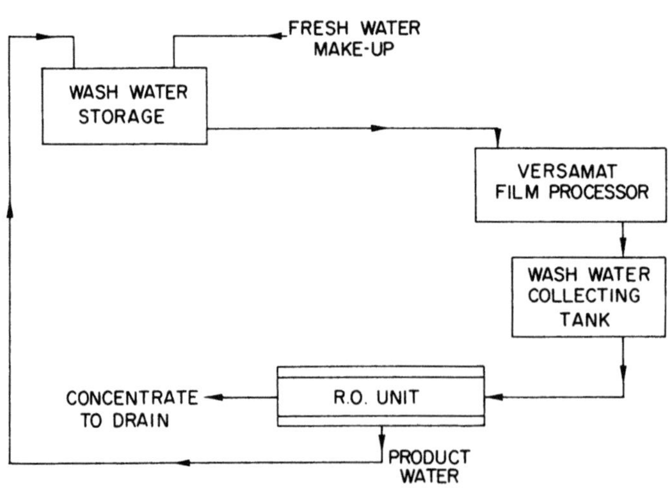

Figure 4. Flow diagram for the reuse of wash water from Kodak Versamat Film Processor.

TABLE 1. REJECTION OF INDIVIDUAL CHEMICALS BY EASTMAN RO-94 MEMBRANE

Test Cell: 1-5/8-inch diameter, 600 psi, 80 F.

Chemical Tested	Concentration In Feed (g/l)	Concentration In Product (g/l)	Percent Rejection	Flux (gal/ft^2/day)
Benzyl Alcohol	2.5 ml/l	2.5 ml/l	0	13
Elon (with Sodium Sulfite)	6.5	5.3	18	10
Hydroquinone (with Sodium Sulfite)	6.0	5.5	9	11
Formalin	27 ml/l	20 ml/l	25	14
Sodium Sulfite	7	0	99+	11
Sodium Thiosulfate, Hydrated	5	0.02[a]	99+	39
	220	0[b]	99+	7
Sodium Ferrocyanide, Decahydrate	20	0	99+	20
Potassium Ferricyanide	50	0	99+	14
	200	1[a,c]	99+	8.5
Sodium Bromide	5	0	99+	20
Potassium Dichromate	20	1.3	94	–
Hydroquinone Monosulfonate	10	1.2	88	–
Silver Thiosulfate Complex (Expressed as Silver)	0.10	<0.001[a]	99+	38

(a) Determination with RO-89 membrane.
(b) Feed Pressure = 1000 psi.
(c) Feed Pressure = 1200 psi.

Recovery of Silver, Fixing–Bath Chemicals, and Water. As film is processed, silver is transferred from the film to the fixing bath, forming a large inorganic complex ion. The removal of silver from the complex in the fixing bath is accomplished in large laboratories by an electrolytic process. However, it is not practical to electrolyze solutions containing less than about 500 mg/l of silver. Wash waters generally run from 10 to 100 mg/l of silver, thus silver that is carried over into a wash water cannot be recovered electrolytically from the wash water. Fortunately, the large ion is easily separated by reverse osmosis. The concentration of the silver in the reverse-osmosis product water is high enough to be recoverable electrolytically. Furthermore, the thiosulfate ion is separated along with the silver complex and it is reusable in the fixer, and the product water may be reused.

To illustrate how reverse osmosis can be utilized to recover silver, a ten-module unit (described in the "Semiplant Equipment" section of this paper) was used to treat the fixer wash water from a process for Ektacolor paper. The results in Table 2 show that when the silver was 30 to 40 mg/l in the feed it was reduced to not over 2 mg/l in the product water and that the silver in the concentrate was high enough for electrolytic recovery. Note that the process recovers about 95 percent of the silver that would otherwise be lost. A tighter reverse-osmosis membrane would probably have proved even more advantageous because of the high dollar value of the silver.

To illustrate how reverse osmosis can be used to recover fixing bath chemicals, a reverse-osmosis semiplant test on a typical processor wash water was demonstrated in our laboratory by a representative of a company that markets reverse-osmosis units. The used wash water in these experiments was that following the fixing baths of a Kodak Versamat film processor. A second test of similar wash water was conducted by an outside company at their location. Results of this second test are shown in Table 3. At all percentages of water recovered, rejection of 99+ percent of sodium thiosulfate was attained. This is in good agreement with the chemical separation reported on the laboratory equipment. An overall solids rejection (essentially salts) of 94.5 percent was obtained at the average of 90 percent water recovery. It is also noted that a silver concentration of 60 mg/l was found in the concentrate with only 0.3 mg/l, or less, in each of the product portions.

Further tests were conducted in our Kodak laboratories using 20 tubular modules with a total membrane area of about 140 ft^2. Tests on 0.5 percent sodium-chloride solution gave 90 percent rejection at 600 psi. Solutions of 1, 5, and 10 g/l of $Na_2S_2O_3$ were run through the reverse-osmosis unit resulting in a 96, 85, and 72-percent water recovery, respectively. A rejection of 97 percent was achieved at each of the three concentrations with a feed flow rate of 4 liters/minute.

To test the concept of reusing reclaimed wash water by the reverse-osmosis process, an open-loop experiment was carried out on a film on Kodak Versamat film processor, Model

TABLE 2. REMOVAL OF SILVER FROM WASH WATER

Run	Feed (mg/l)	Product (mg/l)	Concentrate (mg/l)	Percent Rejection
1	10	1	150	90
2	40	1	500	97.5
3	30	2	1,100	93

11-A. Since the water quality achieved in the previous experiment was better than needed, a more open membrane module stack was used (40-50 percent sodium chloride rejection, 5 g/l NaCl solution, 600 psi operating pressure). The processor/reverse-osmosis flow chart is shown in Figure 4. The unit needed to operate only about 80 percent of the time because it processed the wash water faster than the processor furnished it. The unit was turned off by a float switch located in the water holding tank. Fresh makeup water was added to compensate for the loss by the concentrate throw-away. During the 8-hour experiment, 9 rolls (9-1/2 inches wide x 500 feet long) of Kodak Plus-X Aerographic Film were processed. The feed rate going into the module stack was 2400 ml/min. Figure 5 shows the concentration of sodium thiosulfate before and after passing through the reverse-osmosis unit. Because of the particular membrane chosen, only about 70 percent of the sodium thiosulfate was rejected. During the 8-hour period, 87 percent of the wash water was reclaimed. The chemical concentration of sodium thiosulfate in the wash water and in the reclaimed water reached steady state in about 4 hours. Figure 6 shows the residual sodium thiosulfate in the processed film. The level was well within the acceptable commercial limit of 15 μg per cm^2.

It has been shown to be technically feasible to apply reverse osmosis for the recovery and reuse of fixing-bath chemicals, the recovery of silver, and the reuse of the wash water. The first two benefits minimize water pollution even if the economic benefits may not be sufficient to justify the recovery processes. In areas of insufficient or extremely expensive water supplies, the use of reverse osmosis for recovery of wash water may be the only practical way for a photographic processor to stay in business.

TABLE 3. QUALITY OF PRODUCT WATER FROM FIXING-BATH WASH WATER (400 PSI)

		Analyses		
		Feed, pH	$Na_2S_2O_3 \cdot 5H_2O$ (mg/l)	Silver (mg/l)
Feed		5.3	5,000	
Product Fraction, percent water recovered				
(1)	0-20	5.8	25	0.1
(2)	20-40	5.9	19	<0.1
(3)	40-60	6.1	23	<0.1
(4)	60-80	6.2	25	0.1
(5)	80-85	6.2	27	0.3
(6)	85-90	6.1	28	<0.1
(1) - (6)	Average		24	
Concentrate		5.4	33,200[a]	60

(a) 99+ percent of that in feed.

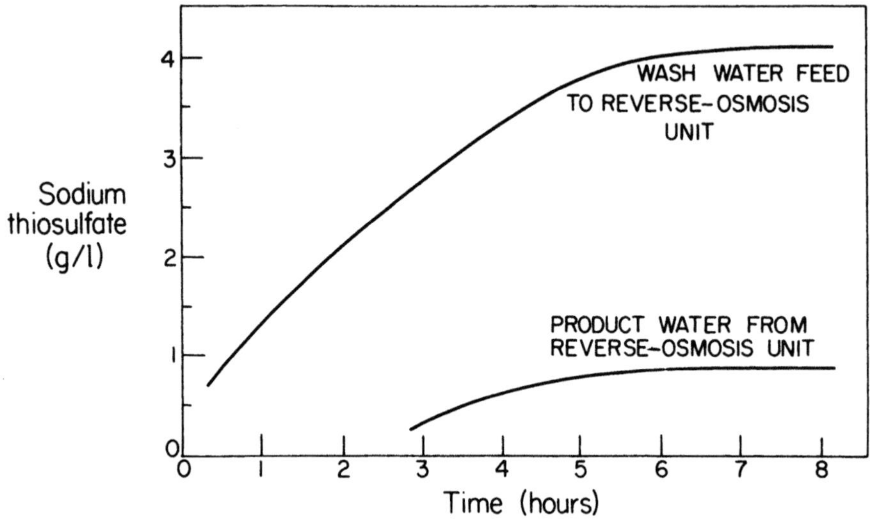

Figure 5. Sodium thiosulfate (hypo) content of wash water before and after reverse-osmosis treatment.

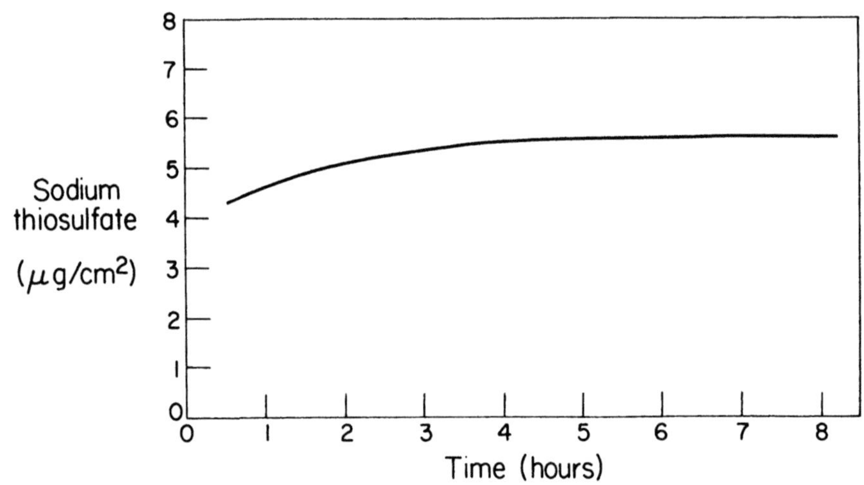

Figure 6. Sodium thiosulfate content of film using wash water from reverse-osmosis unit.

Bleach Recovery. Ferricyanides and ferrocyanides were shown in our initial studies to be rejected almost completely by a "tight membrane". A washing step follows the bleaching in the Kodacolor film C-22 process, and in the Ektaprint C processes. If these chemicals are to be reused to prepare replenishers, concentrations of about 75 g/l of ferro- or ferricyanide salts are necessary.

Using the tubular reverse-osmosis unit (90 percent rejecting membranes) a 15 g/l solution of sodium ferrocyanide was tested. Portions were concentrated to 71 and to 96 g/l at 600 and 700 psi, recovering 91 percent of the water. A rejection of 95.4 percent was obtained at 700 psi with an 89 percent water return. At these concentrations of ferrocyanide the product flow rates dropped about tenfold from the start to the end of the tubular stack. These results indicate the feasibility of recovering bleach from wash waters so that it can be reused. It may be a practical system in an area where the sewer code limits the concentrations of ferricyanide.

Treatment of Wash Waters for Storm Sewer Disposal. Some municipalities have volume restrictions on their sanitary sewer system, but no volume restrictions on their storm sewers. If photographic processors find it necessary, it may be practical to separate their large volumes of wash-water overflows from the small volumes of concentrated-processing overflows in their plant. The wash-water overflow might be passed through reverse-osmosis units to lower the waste load still further so that the product water could be discharged to the storm sewer or reused in certain applications. The small volume of concentrate would have to be disposed of in some other way.

As a severe test of the effectiveness of such a proposal, overall effluents from photographic processing (not just wash waters) were simulated for three photographic processes. These effluents were passed through a 20-module tubular reverse-osmosis unit at 600 psi and at a feed rate of 2.7 liter/minute. The membranes were the 90 percent sodium-chloride-rejecting type.

One measure of the separation of salts from the waste waters is the chemical oxygen demand (COD) before and after it passes through the reverse-osmosis unit. Table 4 shows that the COD was lowered more than 80 percent and that the product effluent contained about 87 percent of the input water. The 13-percent concentrate would have to be disposed of in some acceptable way. Even more favorable separations would be expected using more dilute wash waters rather than overall waste effluents. Other tests in addition to COD would be required for a more complete appraisal of the treated wastes.

Summary of Possible Applications to Photographic Processing

This research study has shown that:

(1) Final wash-water in a photographic process can be put through a reverse-osmosis unit to reclaim 85 to 90 percent of the water for reuse.

(2) Reverse osmosis can be used to concentrate ferricyanide and ferrocyanide in bleach wash water. The concentrate might be used in preparing a bleach replenisher and the water might be used for washing.

TABLE 4. TREATMENT OF SIMULATED WASTE WATER FROM PHOTOGRAPHIC PROCESSES BY REVERSE OSMOSIS

Pressure: 600 psi
Feed Rate: 2680 ml/min
20-Module Unit: 7 sq ft/module

Process	COD (mg/liter) Ektachrome ME-4	Kodachrome K-12	Ektaprint C
Feed	265	280	350
Product	30-45	40	65
Concentrate	1200	2485	4750
Percent COD Rejection	87	86	81
Percent Water Recovery	80	90	94

(3) Over 90 percent of the silver present in photographic fix-bath wash waters can be concentrated to levels where conventional electrolytic equipment can be used for silver recovery.

(4) An overall effluent from photographic processing can be passed through a reverse-osmosis unit to concentrate the oxygen-demanding chemicals into about 15 percent of the original volume.

In conclusion, it is emphasized that the research work reported here has shown only the technical feasibility of applying reverse osmosis to several aspects of photographic processing on a small scale. Whereas the application of reverse osmosis for water reuse, chemical reuse, and pollution control appears attractive, we have as yet made no commercial trials and no economic studies for commercial application. We are not in a position to recommend reverse osmosis for use by photofinishers at this time and, of course, every situation would differ, depending on the local conditions. However, it is felt that the potential benefits justify our continued investigations into possible commercial applications.

REFERENCE

(1) S. Sourirajan, "Characteristics of Porous Cellulose Acetate Membrane for the Separation of Some Organic Substance in Aqueous Solution", *I & E C Product Research and Development,* 4 (3) (September, 1965).

ACKNOWLEDGMENT

The authors wish to acknowledge data received from Mr. A. C. Cooley, Mr. L. I. Edgcomb, Mr. F. L. Halbert, and Mr. C. E. Upham, all of the Photographic Technology Division, Kodak Park.

ULTRAFILTRATION WATER TREATMENT

Clifford V. Smith, Jr., and David Di Gregorio
Dorr-Oliver Incorporated
Stamford, Connecticut

INTRODUCTION

In spite of the increasingly serious problem of pollution in our water supplies, the separation processes used to treat contaminated surface-water supplies have not changed appreciably during the past 30 to 50 years. However, recent developments in ultrafiltration offer public-health engineers a more efficient technique for removing contaminants in water to make it potable.

Ultrafiltration, as used here, is a sieving operation in which membrane pore size is the principal determinant in the separation mechanism. This approach is in contrast to reverse osmosis, in which diffusion is the main mechanism.

Ultrafiltration

The theory of ultrafiltration is not new. What is new is its successful, practical application to the separation of contaminants such as bacteria and organic colloidal impurities.

In the various separation systems listed in Figure 1, the ultrafiltration membranes used are noncellulosic organic polymers having an asymmetric, extremely thin (5-micron) surface layer or skin and a porous substructure of the same material. The total membrane thickness is only 6 to 8 mils; consequently, to handle the 20 to 50 psi used in water treatment, the membranes are reinforced with a nonwoven paper material for added mechanical support. Separations take place at the membrane surface, **not** within the substructure (see Figure 2).

In the tests reported here, the ultrafiltration systems comprise a series of modules, each providing 60 square feet of surface area. (Each module contains three replaceable cartridges, and each cartridge contains 20 square feet of membrane housed in a glass-reinforced polyester

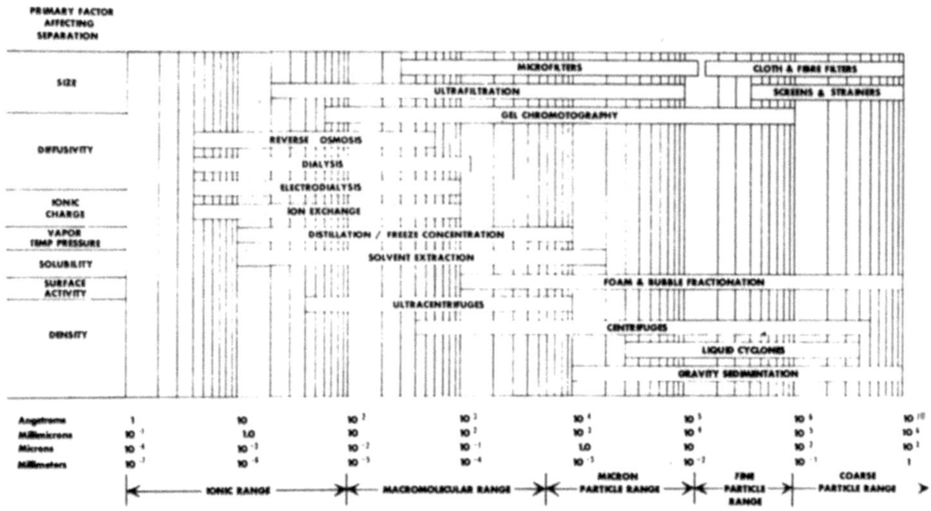

Figure 1. Useful ranges of various separation processes.

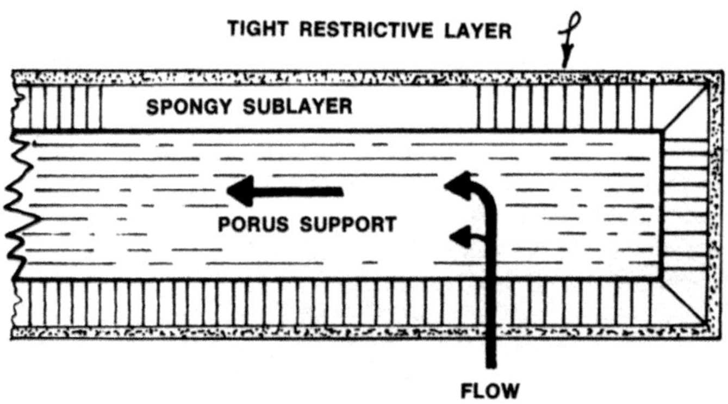

Figure 2. Schematic membrane and support cross section.

casing.) The supported membrane sheets are spaced 1/8 inch apart to allow the system to operate under conditions in which the feed stream has an appreciable amount of suspended material in it (see Figure 3).

Inherent in the use of membranes for phase separation is the fundamental problem of concentration polarization. The separation of one species from another at the surface of a membrane results in local concentration of the rejected species at the membrane surface. This concentration in turn causes an increase in density and viscosity of the laminar sublayer at the surface of the membrane. Associated with these changes is the decrease in membrane flux. The more severe is the polarization, the poorer is the long-term flux characteristics of the membrane.

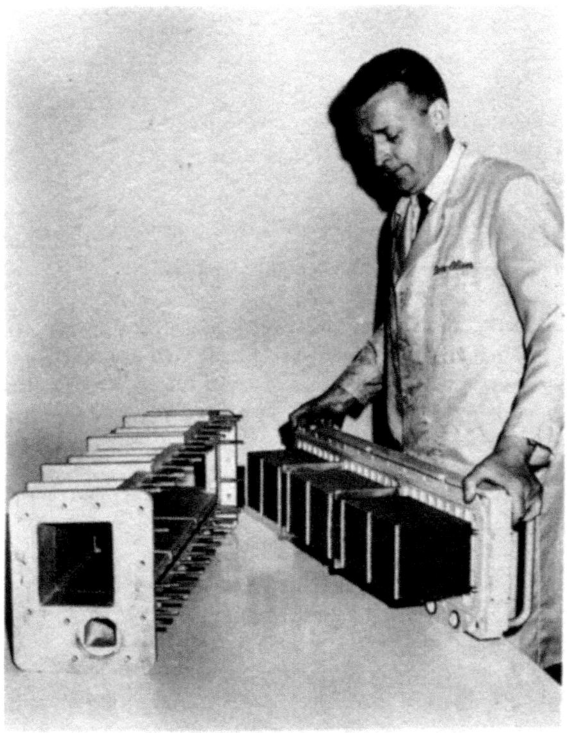

Figure 3. Basic iopor membrane module.

One technique for minimizing polarization is to establish turbulent mixing in the system to reduce the concentration profile along the flow channel. In Dorr-Oliver systems, this turbulence is obtained by recirculating the feed across the membranes at superficial velocities of 4 to 6 ft/sec.

The general term "membrane separation" includes a broad spectrum of transport processes referred to as filtration, ultrafiltration, reverse osmosis, etc. Although a complete theoretical discussion of these transport mechanisms is beyond the scope of this paper, a brief description of the fundamental transport process of ultrafiltration is in order.

The mathematical equation that describes pore flow through a membrane is known as Darcy's Law, which, when modified for osmotic pressure, can be written for solvent and solute flux as

$$J_1 = \frac{K_1 \Delta P - K_3 \Delta\pi}{t}$$

$$J_2 = \frac{K_2 C_1 \Delta P}{t}$$

$$\frac{C_1}{C_0} = \frac{K_1 \Delta P - K_3 \Delta \pi}{K_2 \Delta P}$$

where

J_1 and J_2 are the solvent and solute flux, respectively
K_1, K_2 and K_3 are constants
C_1 is the upstream concentration
C_0 is the downstream concentration
t is the membrane thickness
P and π are the applied and osmotic pressures, respectively.

Since operating conditions in practically every application are far from ideal, the mathematical relationships can only approximate performance of membrane systems. For example, defects in membrane structure and presence of secondary films are two of the factors in making the equation something less than perfect. Our experience with membrane systems indicates that flux is a function of many variables such as pressure, velocity, temperature, and membrane loop concentration.

DORR-OLIVER ULTRAFILTRATION WATER-TREATMENT SYSTEMS

During the past 2 years, Dorr-Oliver has been engaged in a program to develop a membrane ultrafiltration system for treating surface water . . . a system that would yield an effluent which meets the United States Public Health Service and World Health Organization standards and yet be comparatively simple.

Drinking-water standards for the USPHS specify that

(1) When the membrane-filter technique is used, the arithmetic-mean coliform density of all standard samples examined per month shall not exceed 1 per 100 ml.

(2) Coliform colonies per standard sample shall not exceed 3 per 50 ml, 4 per 100 ml, 7 per 200 ml, or 13 per 500 ml in

(a) Two consecutive samples

(b) More than one standard sample when less than 20 are examined per month

(c) More than 5 percent of the standard samples when 20 or more are examined per month.

WHO standards state

(1) Ninety percent of samples examined during a year shall be free from coliform

(2) The MPN index of coliform microorganisms shall be less than 1.0

(3) None of the samples shall have an MPN index of coliform bacteria in excess of 10.

In the case of inorganics, ultrafiltration membranes cannot perform separation unless the inorganics have formed into large molecules such as colloids. Normally, when considering various sources for a proposed public water supply, the optimum source is that which does not require removel of inorganic ions.

The simplicity of the membrane ultrafiltration system can be appreciated when comparing it with a conventional water-treatment system. This comparison of flowsheets is shown in Figure 4.

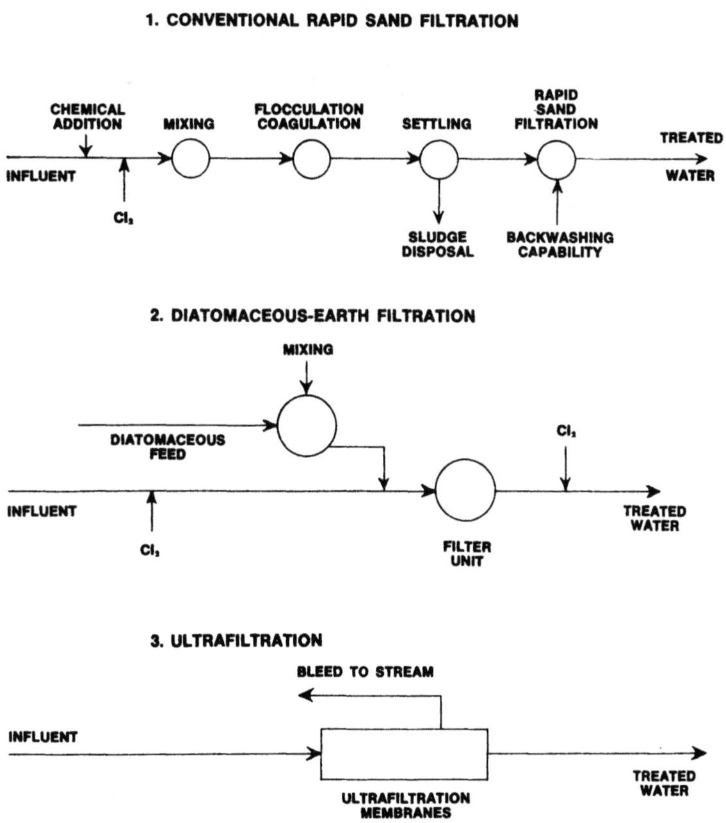

Figure 4. Schematic flowsheets of various water-treatment systems.

In addition to offering a simple, one-step automatic process, the ultrafiltration membrane provides a positive barrier between the contaminated supply and the treated water. On the other hand, the diatomaceous-earth system, while it does provide an emergency bypass, may, in case of power failure, be susceptible to contamination of the treated water.

Although conventional rapid sand filtration provides several lines of defense, chlorination of the treated water has been imperative when treating grossly polluted supplies.

Increasing concern over transmission of virus diseases prompted an earlier study of the ability of a membrane possessing substantial flux to retain virus. As a part of this study, an upstream system was inoculated with E. coli phage in water suspensions containing 10^7 to 10^9 organisms per ml.

The results of this experiment are shown in Table 1. Although E. coli phage have no sanitary significance, their size and ease of quantitative measurement suggested their use.

Table 2 compares E. coli phage with several significant sanitary viruses.

To demonstrate the effectiveness of ultrafiltration membranes in treating an actual water supply, a 500-gpd pilot plant was installed near the Mianus River in Greenwich, Connecticut. This plant is shown in Figure 5 and schematically in Figure 6.

Pilot-plant operation demonstrated that substantial fluxes can be obtained with ultrafiltration membranes. No chemicals and very little attention were required.

To demonstrate coliform rejection, their concentration in the effluent was measured during a 2-month period. The following results were obtained:

Coliform Organisms/100 ml	Percent Time Equal to or Less Than
0	50
2	83
2.2	90
6.2	97

The positive results were probably caused by contamination during sampling.

A 20-consecutive-day test run yielded the following color and turbidity values:

Sample	Color	Turbidity
Feed	32	4.8
Effluent	<5	0

Figure 7 represents fluxes obtained during a 60-day test. Effluent color during this run averaged less than 5 units and turbidity was less than 0.2 JTU.

On several occasions, due to loop bleed failure, the color of the membrane loop contents rose over 700 units without affecting the color or turbidity of the effluent.

TABLE 1. RESULTS OF VIRUS RETENTION STUDIES[a]

System pressure was 50 psi.

Date	Elapsed Time	Upstream (org/ml)	Downstream (org/ml)	Flux (gfd)
12/20	0	Tap water only		60 - 360
12/21	1	Reservoir inoculated		40 - 80
12/22	2	3.5×10^4	0	15 - 30
		New Membrane Installed		
12/29	0	Tap water only		24, 29, 75
		Reservoir inoculated		
12/30	1	--	--	25, 26, 58
1/3	5	4.7×10^4	0	23, 24, 30
1/4	6	--	--	22, 22, 24
1/5	7	--	--	17, 19, 22
1/7	9	70	0	12, 13, 19

(a) The upstream reservoir was inoculated with viral suspensions containing 10^7 - 10^9 org/ml. Data are summary of results from several Dorr-Oliver membranes.

TABLE 2. APPROXIMATE SIZE OF REPRESENTATIVE VIRUSES

Virus	Size, angstroms
Lymphogranuloma	3,000 (round)
Trachoma	3,000 (round)
Smallpox	2,300 x 3,000
Ectromelia (mice)	2,300 x 3,000
Mumps	1,700 (round)
Influenza	1,000 (round)
E. Coli	650 x 950
Encephalitus	400 (round)
Yellow Fever	400 (round)
Poliomyelitis	300 (round)

Figure 5. Iopor ultrafiltration water-treatment plant, 5000 GPD capacity.

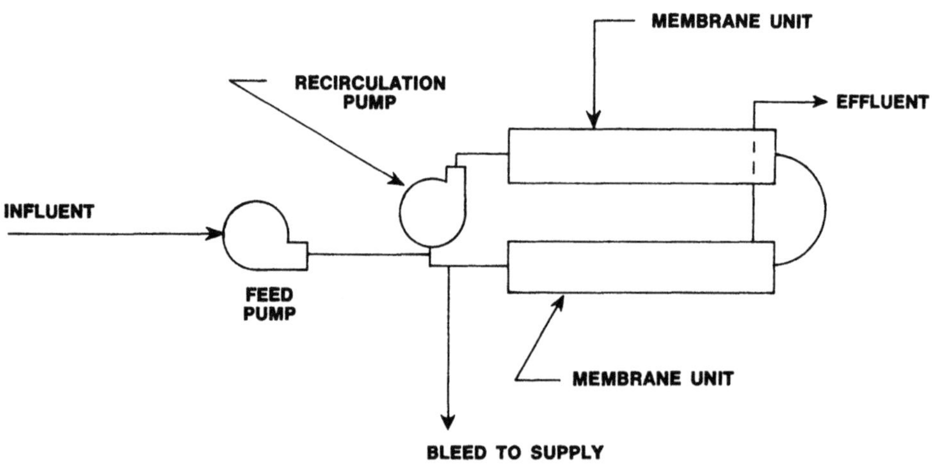

Figure 6. Schematic diagram of ultrafiltration water-treatment system.

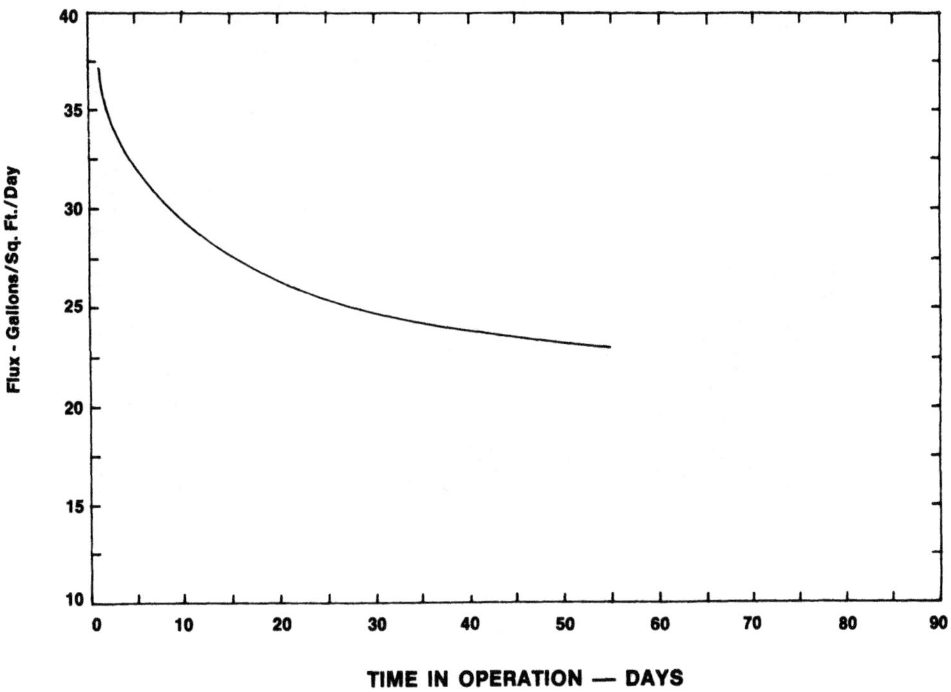

Figure 7. Average flux at ultrafiltration water plant.

It became obvious during operation of the pilot plant that several operating parameters strongly influence the long-term flux patterns. These factors include system pressure, velocity, and membrane loop concentration. Velocity is important, because it is responsible for maintaining a clean membrane surface and for preventing excessive concentration polarization.

To obtain desirable flux patterns, the relationship between pressure, velocity, flux, etc., should be known, and the operating system should function in consideration of this relationship.

An experimental program was developed to determine the relationship between the above-mentioned operating parameters. A smaller experimental pilot plant utilizing 1-square-foot membranes was constructed and installed at the location described above.

Four independent membrane loops are used in the experimental system to realize the desired flexibility (see Figure 8). With the smaller pilot plant, all operating variables may be held constant, except for one which may be varied between the four independent loops. This program is now in progress.

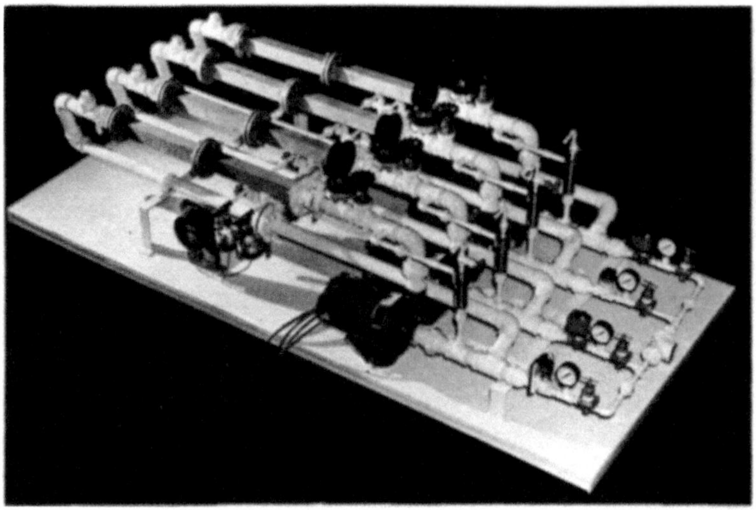

Figure 8. Experimental Iopor filtration pilot plant.

Economics

Current and projected costs for the membrane ultrafiltration system are shown in Figure 9. Based on present costs, membrane water-treatment systems can be marketed for $20 to $50 per square foot, depending on plant capacity. Operating costs range from $0.65 to $1.25 per 1,000 gallons.

Our experience indicates that water costs for conventional water-treatment systems as well as for membrane systems vary considerably from country to country and even within countries. For conventional systems, the variation is due primarily to differences in chemical and power costs, while in membrane systems the variation is due to differences in power only. For instance, it was found that power costs in Latin America ranged from 2 to 10 cents/kwhr.

Continuing Research and Outlook

Dorr-Oliver is continuing a program of research and development to demonstrate the efficiency of membrane ultrafiltration in treating surface-water supplies.

In addition to the pilot plants now being operated, other test locations in Europe and Latin America as well as in the United States are under consideration.

As for general acceptance of the membrane system, there are two basic factors involved . . . technical and economic. Technically, the system is capable of meeting the requirements of international and local health ordinances. Economically, it costs more than a conventional chemical-coagulation/sand-filtration/chlorination system.

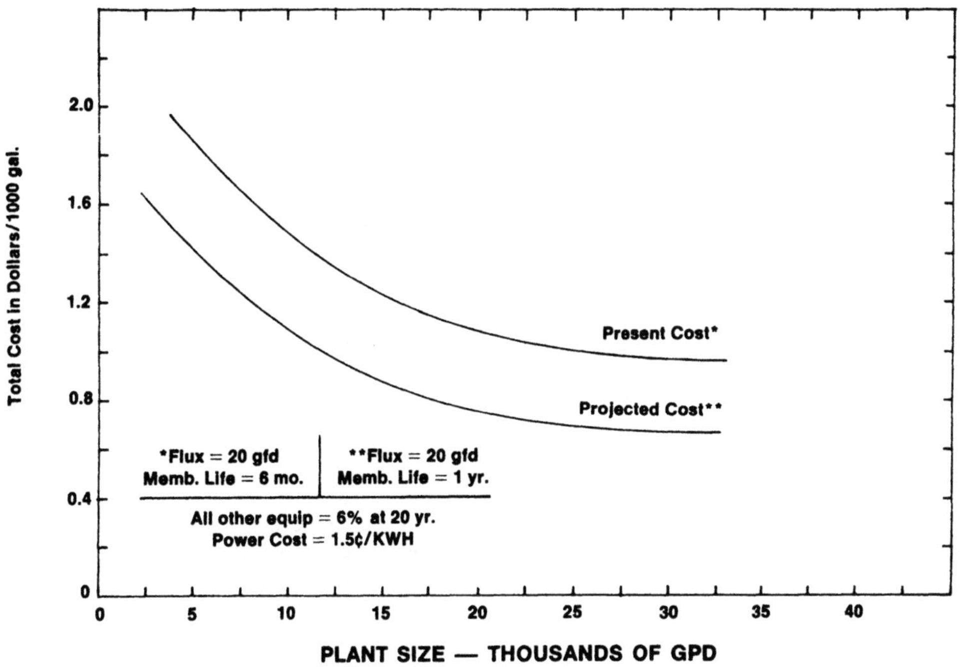

Figure 9. Total cost for membrane water treatment.

However, the cost for any social or health service (in this case, superior drinking water) cannot be measured strictly in dollars and cents by capital costs or cost per gallon. The factor determining acceptance is not the price paid for but the value of the service, and the value placed upon anything is closely related to the culture, the standard of living, and the priorities prevailing in a given community.

REVERSE OSMOSIS: APPLICATION TO POTATO-STARCH FACTORY WASTE EFFLUENTS

W. L. Porter, J. Siciliano, S. Krulick, and E. G. Heisler
Eastern Utilization Research and Development Division
Agricultural Research Service
United States Department of Agriculture
Philadelphia, Pennsylvania

One of the major problems of our country, and most others, is pollution of the water supplies. In this paper, we hope to describe for you a specific pollution problem of great importance, a possible means of alleviating this problem, and a possible role of reverse osmosis in the technology under consideration.

Since all of you are well versed in most aspects of reverse osmosis, little time need be spent on background material for this phase. It is assumed, however, that few of you are familiar with the problems of pollution and waste treatment faced by the food industry in general and the potato-starch industry in particular. Therefore, to make it easier later, a little background material will be presented. Only problems concerning the potato-starch industry will be discussed.

In the manufacture of potato starch, the tubers are removed from storage by means of flowing-water flumes, which transport them to a conveyer and at the same time remove vegetative parts, soil, and stones. This water is usually sent to settling ponds prior to recycling to the flumes and thus does not enter into the immediate problem. After washing, the tubers are ground or rasped to a slurry. This slurry is screened and washed by various means with large quantities of water to separate the starch from the fibrous material, and, finally, the starch is centrifuged and again washed with copious quantities of water to remove all remaining soluble materials. On the average, the solids content of potatoes is 19 percent (Table 1). The starch content runs about 13 percent on a fresh-weight basis. This means that about 6 percent of the fresh weight consists of nonstarch solids, and these end up in the waste effluent. Until about 10 years ago, disposal of this waste material consisted simply of its being dumped into the nearest river. Esthetic and survival considerations, however, have made it necessary to end pollution of this nature, and Federal and state laws have been enacted which require treatment of the effluent before returning the water to the river.

Considering an average starch plant producing 30 tons of starch per day, about 14 tons of nonstarch solids end up in the waste effluent from the factory. This waste contains both insoluble materials, termed primary waste, and soluble materials, termed secondary waste. The

TABLE 1. AVERAGE COMPOSITION OF POTATOES PROCESSED IN MAINE STARCH FACTORIES[(a)]

Substance	Percent
Starch	13
Protein (N x 6.25)	2
Cellulosic material	1.5
Sugars	0.5
Minerals (ash)	1
Miscellaneous minor constituents (total)	1
Water	81

(a) From *Potato Processing*, Talburt and Smith, AVI Pub. Co., p 451 (1967).

starch industry has developed settling, screening, centrifuging, and filtering procedures designed to remove the primary waste; to date, disposal of this product has consisted of selling it as a cattle feed. Therefore, it does not loom as an immediate pollution problem.

The secondary waste is another story. This material contains protein, free amino acids, organic acids, sugars, inorganic ions, and other compounds in minor concentrations. The output of this waste from a 30-ton-per-day starch plant is about 300 gpm or about 432,000 gallons per day. The solids content is about 0.5 to 1.0 percent, and it has a chemical-oxidation-demand (COD) requirement of approximately 9000-14000 mg per liter. This is equivalent to a city of approximately 85,000 people, with respect to waste-disposal potential. Presently, it is pumped into lagoons, where biological action removes the dissolved organic solids. Lagooning is a relatively inefficient process and, since the potato solids are quite resistant to treatment, it is necessary to employ large land areas for the treatment process. Even with this treatment, only about 80 percent of the COD requirement can be removed. In addition, the odor emanating from many treatment ponds is most unpleasant, to say the least. In the near future, it is quite possible that the law will require treatment of such waste to the extent that the renovated water returned to the river will be no different than the water that was removed up river for use in processing. It has been estimated that the facilities required to produce this quality of water effluent would cost 80 percent or more of the present plant investment. Since starch factories are marginal enterprises, they could not afford such expenditures and would have to go out of business, thus causing a shortage of this important item and also loss of an outlet for cull potatoes that is very important to the potato industry. Since potato starch is a valued product in many applications with the industry volume at approximately $10,000,000 per year based upon the selling price, this would be a serious loss.

The present methods of renovation do not bring any monetary return on the secondary-waste-treatment investment. Improving this situation may be the means for solution of the dilemma. Any treatment procedure that includes processes yielding a monetary return would ease and distribute the cost. Such return could mean the difference between failure and continued business for a marginal company.

The proteins and amino acids have been shown to contain more than adequate amounts of lysine and methionine for human and animal nutrition[(1)] and, as such, would have considerable value on the market as additives to certain grain products and as starting materials for new food products or flavoring agents. The Dutch have developed a process for recovering proteins[2], and the USDA has published on means of recovery of the free amino acids by ion exchange.[3,4] The inorganic ions must be removed prior to the ion-exchange removal of free amino acids[5], and they have value as a fertilizer because of their high potassium content. The organic acids have value due to their high citric-acid content.[6] A project aimed at recovery of these products has been initiated. By removing these organic compounds, the COD requirement of the remaining effluent would be decreased to 20-25 percent of that of the original waste. Also, the organic-acid recovery step removes 98 percent of the phosphorus, which would help to eliminate algae growth in the final lagoon treatment where the remaining organic matter, mostly sugars, would be biologically treated. The income from the sale of these products would probably pay a major portion of the cost of treatment.

At a solids concentration of 0.5 percent, the efficiency of recovery of protein by heat and acid coagulation and the removal of amino acids, inorganic ions, and organic acids by ion-exchange procedures is relatively low. At a solids concentration of 2-4 percent, these processes have been shown to be quite efficient.[5] The usual concentration methods, such as distillation and freezing, are expensive to use, especially when such large volumes are involved. Therefore, studies are in progress on the use of reverse osmosis for application to these problems because the actual expense of operation is low. This latter statement will be discussed more fully subsequently.

For several reasons, the Havens "Osmotik Processor"* was chosen for this work (Figure 1). This apparatus is portable and has an input capacity of about 3 gpm, which is a good size for development work in which limited supplies of the solution to be concentrated are available. In addition, the modules are made up of 1/2-inch-diameter tubes and serve this purpose extremely well because the starch-waste influent usually contains some finely divided, insoluble solids. These solids tend to settle on a flat membrane or to plug up a module made in rolls or disks with narrow spacers. The tube type can be easily opened and flushed with water or scoured by the use of a 1/2-inch-diameter polyfoam plug followed by water. Sterilization of the apparatus with hypochlorite is facilitated by the geometry of the module.

The apparatus, as purchased, had 24 modules of seven tubes each. Eight modules contained the coarse-porosity membrane, eight contained the membrane of intermediate porosity, and eight contained the fine membrane. They can be operated individually, in parallel, or in series. Each unit of eight modules contains 52 square feet of membrane. Due to a limited supply of waste effluent, much of the early work was performed by returning the concentrate to the supply. This, of course, produced an ever increasing concentration of effluent.

Most of the work to be discussed was done on a simulated waste effluent produced by grinding potatoes, filtering, and washing to give a concentrated product (in a manner similar to that employed in a starch factory) which was diluted to the desired solids content before use in a particular experiment.

Due to the limited time available, a few typical examples of experiments have been chosen which indicate the possibilities of the procedure. A preliminary run, using all modules in series (coarse, intermediate, and fine, in that order), an input of 50 gallons of the simulated

*Mention of company or trade names does not imply endorsement by the Department over others not named.

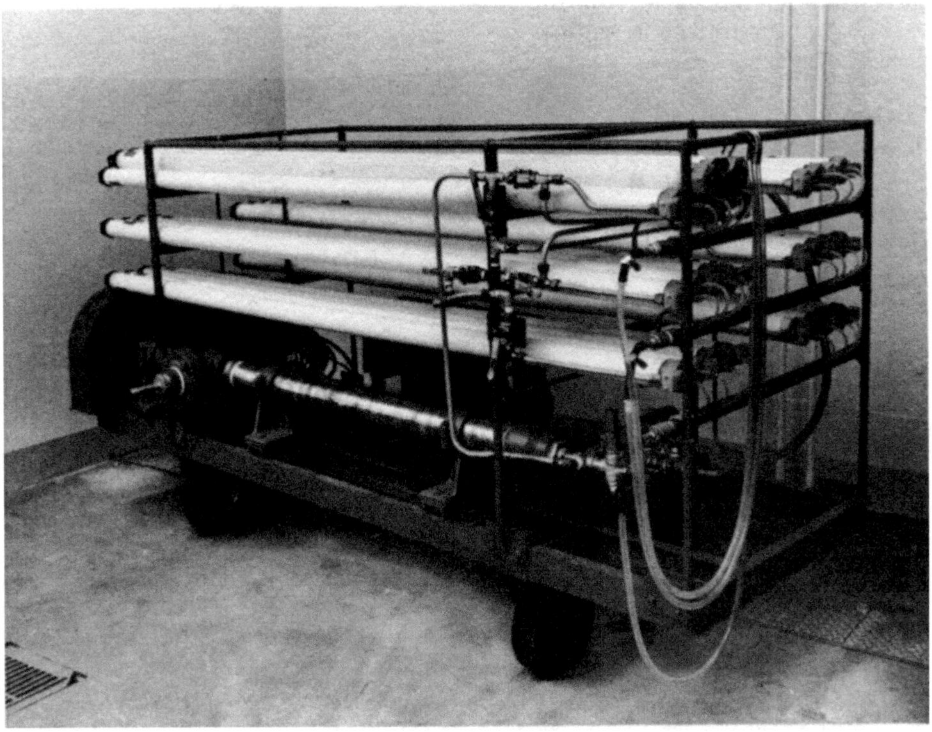

Figure 1. Experimental reverse-osmosis equipment.

waste effluent containing 0.32 percent solids, a pressure of 600 psi, and recirculation of the concentrate, produced in a period of 45 minutes the results shown in Table 2.

This short experiment showed that the waste effluent could be concentrated by reverse osmosis. However, little, if any, information was obtained for individual membrane types as to flux, comparison of membrane properties, solids in the permeate, etc.

Using a concentrating system in which the permeate was removed and the concentrate was returned to the feed tank to be recirculated, experiments were run by pumping a measured volume of waste water through the coarse-membrane modules only, the intermediate-membrane modules only, and the fine-porosity membranes only. All permeate was collected and, at the end of each run, all the concentrate was collected. Using this method, a reasonable measure of all volumes including initial feed, permeate, and concentrate, was obtained. Analyses of these three fractions gave a measure of the performance of each system with respect to retention of the waste-water constituents in which we are interested.

In a typical experiment, about 55 gallons of the waste water, at 0.5 percent solids concentration, was charged to a feed tank. This was pumped through the medium or intermediate-porosity-membrane modules with the concentrate being recirculated. All permeate was collected. The operating pressure was 600 psi and the pumping rate was 3 gpm. When 41 gallons of permeate had been collected (about 75 percent removal of water), the run was

TABLE 2. CONCENTRATION OF WATER USING VARIOUS REVERSE-OSMOSIS MODULES IN SERIES

Elapsed Time, min	Membranes Used	Concentrate Solids, ppm[a]	Permeate Solids, ppm	Permeate Conductivity, ppm Cl⁻
10	All	4790	87	
13	Coarse	5050	148	80
22	All	6090	99	
29	Fine	6680	25	10
30	Intermediate	--	39	20
35	All	8400	110	
39	All	9700	137	
45 (end)		9620 (0.96%)[b]		

(a) Concentrate sample taken when concentrate left apparatus but before mixing with remaining concentrate to be recycled.

(b) Final solids in concentrate measured after thorough mixing of entire concentrate.

stopped. All of the final concentrate was collected. The total solids of 1.9 percent indicated a fourfold concentration At the start of the run, the flux was 10.2 gf^2d at 77 F. At the end, the flux was 8.8 gf^2d at 95 F. The pH of the concentrate changed from 6.4 to 6.7. A summary of the data is shown in Table 3.

After completing the tests on all three membranes, the following conclusions could be drawn:

(1) Analysis of the permeates from all three membranes showed that the recovered water was pure enough for reuse in a plant. The medium-porosity membrane did a much better job of retaining the desirable constituents than did the coarse one. Solids retention was not greatly improved by the use of the fine membrane.

(2) The average fluxes for the membranes were 11.5 gf^2d for the coarse, 9.5 gf^2d for the medium, and 7.8 gf^2d for the fine.

(3) A fourfold increase in total solids of the concentrate increased the COD by roughly fourfold in all three cases. The COD of the three permeates, however, showed considerable difference. The COD of the coarse-membrane permeate equalled 268 ppm or a reduction of about 94 percent. That of the medium membrane was equivalent of a 98 percent reduction and that of the fine at 40 ppm equalled a 99 percent reduction.

These results indicated that the best choice of the three types of membranes tested was the one of medium porosity. This conclusion was based upon the relationships between flux, retention of desirable waste-water constituents, and reduction of COD.

Most of the membrane designs described in the literature are aimed at increasing the efficiency of production of pure product water rather than the concentrate. Several attempts to increase the rate of concentration, with final automation in mind, have been made. Preliminary experimentation has shown it quite possible to modify the equipment so that a concentrate

TABLE 3. COD OF WASTE WATER ON CONCENTRATION BY REVERSE OSMOSIS

Feed	Total Solids, percent	COD, ppm	
Coarse membrane	0.43	3,670	
Medium membrane	0.49	4,269	
Fine membrane	0.43	3,838	
Concentrate			
Coarse membrane	1.60	13,364	
Medium membrane	1.93	16,163	
Fine membrane	1.57	13,784	
Permeate			
Coarse membrane	0.035	268	[6.4% of ≡ feed COD]
Medium membrane	0.012	91	[2.1% of ≡ feed COD]
Fine membrane	0.005	40	[1.0% of ≡ feed COD]

containing a constant-specific-solids concentration can be produced. The concentrate is returned to the RO unit by-passing the feed tank. As permeate is taken off, an equal volume of fresh feed is introduced. When the desired equilibrium concentration is reached, concentrate is also drawn off at a constant rate and replaced with fresh feed as required to keep up the flow rate. Figure 2 roughly shows the pathways employed in the batchwise configuration and in the continuous configuration. Table 4 shows an example of the continuous results obtained. Table 5 shows the data obtained when the batchwise system was used.

Comparing the times of operation, the flux, and the concentrations obtained, it is apparent, under the operating conditions employed, that the batchwise operation is more efficient. With use of conductivity measurements to follow the concentration by the batchwise procedure, switching from one feed tank to a new one would be quite simple. These data also indicate that other ways for increasing the efficiency should be investigated and experiments are now under way to study parallel and series operation as well as combinations of these to determine the best approach.

These experiments also show the importance of changes in flux as an experiment progresses and as a membrane is subjected to extended use. It is apparent that the flux decreased over the period of time required to carry out the studies reported. No attempt has been made to determine how permanent these changes may be, but such studies are planned for the near future.

Current work, in progress, involves a variation of the type of treatment just discussed. A new membrane is available having a porosity such that the amino acids, organic acids, sugars,

TABLE 4. CONTINUOUS OPERATION OF REVERSE-OSMOSIS EQUIPMENT[a]

Elapsed Time, min	Takeoff	Flux[b], gf^2d	Total Solids, concentrate percent	Permeate Conductance, mhos
0	Permeate	9.4	--	
15	"	8.8	--	2.16×10^{-4}
60	"	6.7	--	3.70
90	"	5.9	--	4.44
120	"	5.3	--	5.14
150	"	4.9	--	5.43
		Concentrate Takeoff Started		
160	"	4.7	3.8	5.38×10^{-4}
190	"	4.7	3.2	4.45
220	"	4.4	2.8	4.25
250	"	3.8	2.4	3.89
310	"	4.2	2.2	3.77
370	"	4.0	2.3	4.19

Total waste treated = 85 gal at 0.5 percent
Total concentrate = 15 gal at 2.7 percent
Total permeate = 70 gal

(a) Medium-porosity membrane used. Area = 52 sq ft.
(b) Pressure = 600 psi.

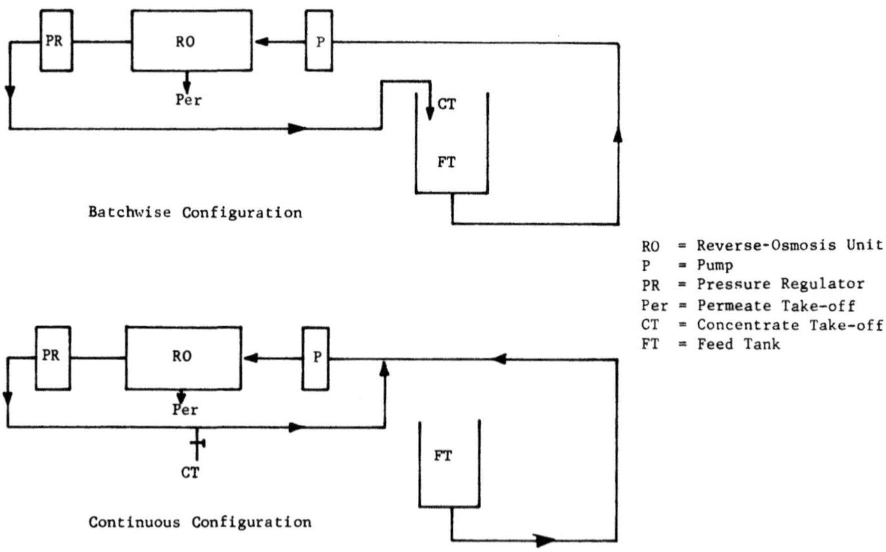

Figure 2. Illustration of batchwise and continuous configurations employed in efficiency experiments.

TABLE 5. BATCHWISE OPERATION OF REVERSE-OSMOSIS EQUIPMENT[a]

Elapsed Time, min	Takeoff	Flux[b], gf2d	Total Solids, concentrate percent	Conductance, mhos
	Concentrate		0.50	
0	Permeate	9.8		
	Concentrate		0.55	1.50×10^{-3}
18	Permeate	9.2		1.14×10^{-4}
	Concentrate		0.58	1.85×10^{-3}
38	Permeate	8.5		1.37×10^{-4}
	Concentrate		0.67	2.08×10^{-3}
58	Permeate	8.1		1.41×10^{-4}
	Concentrate		0.75	2.23×10^{-3}
78	Permeate	7.4		1.54×10^{-4}
	Concentrate		0.91	2.78×10^{-3}
98	Permeate	6.8		1.64×10^{-4}
	Concentrate		1.09	3.35×10^{-3}
118	Permeate	6.7		1.96×10^{-4}
	Concentrate		1.29	4.15×10^{-3}
138	Permeate	6.4		2.51×10^{-4}
	Concentrate		1.62	5.13×10^{-3}
158	Permeate	6.0		3.54×10^{-4}
	Concentrate		3.36	9.87×10^{-3}
168	Permeate	5.1		5.09×10^{-4}

Total waste treated = 55 gal at 0.5 percent
Total concentrate = 6 gal at 3.6 percent
Total permeate = 47 gal

(a) Medium-porosity membrane. Area = 52 sq ft.
(b) Pressure = 600 psi.

and inorganic ions pass through in the permeate and the protein remains in the concentrate. Preliminary trials with this membrane indicate that a multiple-stage concentration could be employed. One of the questions that immediately came up was the sequence of use of the two types of membranes. Figure 3 shows two possible configurations. Each has certain theoretical advantages and disadvantages that must be studied. In the first system, the permeate from the very coarse membrane would contain the soluble, low-molecular-weight materials and the concentrate would contain the protein. The permeate then would pass through the intermediate membrane, producing pure water in the permeate, and the concentrate would contain only the low-molecular-weight solubles. In the second system, the effluent would be passed through the intermediate-porosity membrane to produce pure water as the permeate, and the concentrate, containing all soluble materials from the waste, would be sent through the new, very coarse membrane. The amino acids, organic acids, sugars, and inorganic ions would be in the permeate and the proteins would be in the concentrate.

The first method would pass the entire effluent through the membrane with the highest flux, allowing a smaller quantity to pass through the more dense membrane. Whether the increased concentration of this concentrate would have such a high osmotic pressure as to cause trouble in the finer membrane is not yet answered. The second method may be slower, but the increase in osmotic pressure would not be as important. In either case, the supernatant liquid from the protein precipitation would probably have to be returned to the

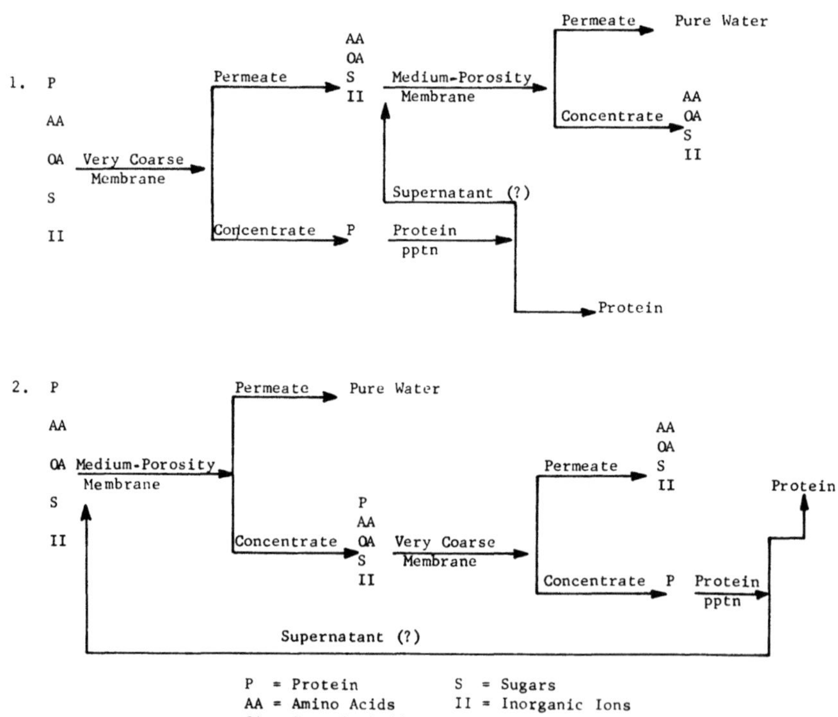

Figure 3. Variations in reverse-osmosis treatments of waste water.

intermediate-porosity-membrane cycle, because the amount of solubles not precipitated by heat and acid coagulation of the protein could be too high to be returned directly to waste or be reused in the plant. The concentration of the desirable constituents obtained by each method will have to be studied, and this may play an important part in the final selection of the method. The fact that the protein can be separated from the other constituents being isolated is important, because the rate of operation of the ion-exchange section would not be dependent on the rate of removal of protein.

This experimental work is either in progress or planned for the near future. Not enough data are available to give direct answers to all of the practical and theoretical questions. However, the final use of the procedure in a plant will be based upon the relative efficiencies of the different possible configurations as well as how they adapt to the recovery processes for which concentration is necessary. The questions, though, serve to point out and demonstrate the problems involved and the types of investigations that must be carried out in practical-application work of this type. It is believed that these and other experiments discussed herein show that the concentration of potato-starch-factory secondary waste can be carried out by reverse osmosis and that actual operating costs would be favorable. As yet, membrane life under these and factory conditions has not been determined. However, we must be practical in any application such as this. At a flux of even as high as 9.5 gf^2d for the intermediate-porosity membrane, treatment of 432,000 gallons per day of waste effluent to produce a fourfold increase in concentration would require the use of 2043 modules (18 tubes each) which, at the current price of $150 each, would cost $306,450 just for the membranes alone. This does not consider replacement of modules due to possible malfunction or the life expectancy being exceeded. Also, no real estimate of what the remainder of a plant, such as pumps, valves, piping, measuring devices, etc., would require is available at this time; but it is safe to say that, with this additional cost, the processing system would not be considered seriously by the industry. We believe the process to have merit and, if the cost of modules can be brought down to a more reasonable figure, it would be employed by the starch industry. There is also a great possibility that the potato-processing industry, made up of manufacturers of flakes, granules, dice, chips, French fries, etc., would also be open to its use, especially for in-plant treatment of waste for immediate reuse of water.

Another phase of the work which will require attention is the sterility problem. As you know, this is a major problem of any food or drug processing industry. As feasibility studies are completed, it will be necessary to demonstrate adequate biological control.

It is hoped that this discussion has shown the type of work required in any practical-application study of reverse osmosis. Since the original cost and replacement cost of membranes is so important to acceptance of the technique by the potato-starch industry, it is also hoped that this paper will help in promoting accelerated studies on membrane production aimed at decreasing these costs.

REFERENCES

(1) T. J. Fitzpatrick, R. V. Akeley, J. W. White, Jr., and W. L. Porter, "Protein, Nonprotein and Total Nitrogen in Seedlings of Potatoes", *Amer. Potato J.*, **46**, 273-286 (1969).

(2) M. Vlasblom and H. Peters, "Recovery of Proteins From Potato Wash Water", Patent No. 87,150, The Netherlands, Jan. 15, 1958.

(3) E. G. Heisler, J. Siciliano, R. H. Treadway, and C. F. Woodward, "Recovery of Free Amino Compounds From Potato Starch Processing Water by Use of Ion Exchange", *Amer. Potato J.*, **36**, 1-11 (1959).

(4) E. G. Heisler, J. Siciliano, R. H. Treadway, and C. F. Woodward, "Recovery of Free Amino Compounds From Potato Starch Processing Water by Use of Ion Exchange. II. Large-Scale Laboratory Experimentation", *Amer. Potato J.*, **39**, 78-82 (1962).

(5) E. G. Heisler, S. Krulick, J. Siciliano, W. L. Porter, and J. W. White, Jr., "Potato Starch Factory Waste Effluents. I. Removal of Potassium and Other Inorganic Cations", Presented at the 19th National Potato Utilization Conference, Big Rapids, Mich., July 27-28, 1969, *Amer. Potato J.*, submitted for publication.

(6) J. H. Schwartz, R. B. Greenspun, and W. L. Porter, "Chemical Composition of Potatoes. V. Further Studies on the Relationship of Organic Acid Concentrations to Specific Gravity and Storage Time", *Amer. Potato J.*, **43**, 361-366 (1966).

SUBJECT INDEX

GPSR Compliance

The European Union's (EU) General Product Safety Regulation (GPSR) is a set of rules that requires consumer products to be safe and our obligations to ensure this.

If you have any concerns about our products, you can contact us on ProductSafety@springernature.com

In case Publisher is established outside the EU, the EU authorized representative is:

Springer Nature Customer Service Center GmbH
Europaplatz 3
69115 Heidelberg, Germany

Batch number: 10684222

Printed by Printforce, the Netherlands